Mission Statement

Napa Sustainable Winegrowing Group (NSWG)

The mission of the Napa Sustainable Winegrowing Group is to identify and promote winegrowing practices that are economically viable, socially responsible and environmentally sound. Specifically, the Napa Sustainable Winegrowing Group promotes viticultural land stewardship through educational outreach to:

- optimize ecological stability and winegrape productivity and quality by understanding and emulating natural processes such as biodiversity, carbon and nutrient cycling, and plant-oil interactions.

- reduce pesticide inputs through cultural practices, biological control and use of alternative materials.

- promote soil health through erosion control, reduced tillage, soil analysis and the amendment of soils with cover crops and compost.

- enhance returns on investment by promoting the value-added nature of sustainable winegrapes along with terroir and increased vineyard longevity.

Government Code Section 7550. Preparation by nonemployees of state or local agency; inclusion of contract and subcontract numbers and dollar amounts

(a) Any document or written report prepared for or under the direction of a state or local agency, which is prepared in whole or in part by nonemployees of such agency, shall contain the numbers and dollar amounts of all contracts and subcontracts relating to the preparation of such document or written report; provided, however, that the total cost for work performed by nonemployees of the agency exceeds five thousands dollars ($5,000). The contract and subcontract numbers and dollar amounts shall be contained in a separate section of such document or written report.

(b) When multiple documents or written reports are the subject or product of the contract, the disclosure section may also contain a statement indicating that the total contract amount represents compensation for multiple documents or written reports. (Formerly Government Code Section 7800, added by Stats. 1979, c.486, p.1655, Government Code Section 1. Renumbered Government Code Section 7550 and amended by Stats.1980, c.676, p.1917, Government Code Section 102.)

Vineyards in the Watershed

Vineyards in the Watershed

Sustainable Winegrowing in Napa County

by Juliane Poirier Locke

Published by the
Napa Sustainable Winegrowing Group
Napa, 2002

Copyright © 2002 by Juliane Poirier Locke

All rights reserved. No part of this book may be reproduced in any form without written permission from the publisher.

Library of Congress Control Number: 2001   129343
       Locke, Juliane Poirier
       Vineyards in the Watershed / Juliane Poirier Locke

       First edition
       Includes index
       ISBN 0-9716622-0-7

Book and cover design, production and typesetting by Laura Lamar, and maps by Max Seabaugh, of MAX Design Studio, Middletown, CA. Produced on the Macintosh using Quark XPress software and Adobe AGaramond and Franklin type font families. Output at dmax *imaging* and printed by Phelps Schaefer Litho-Graphics.
Cover photograph by Astrid Bock-Foster (color); background aerial photo courtesy of Napa County RCD. Copy editing by Tony Bogar.

Napa Sustainable Winegrowing Group
1303 Jefferson Street
Suite 500B
Napa, California 94559

Printed in U.S.A. on recycled paper with soy based inks

For Jordan —

May you love and protect the land

Contents

14   **Preface**

21   **Introduction**
The Watershed

31   **Chapter One: Erosion**
Curbing losses of soil, steelhead and public confidence

43   **Chapter Two: Water**
Conserving a finite supply

61   **Chapter Three: Weeds**
Eliminating high-risk herbicides

73   **Chapter Four: Wildlife**
Protecting habitat, courting diversity on the farm

87   **Chapter Five: Cover Crops**
Preserving and improving the soil

99   **Chapter Six: Vine Health**
Observing changes, integrating practices

119   **Chapter Seven: Organic Farming**
Profitability and land ethics

133   **Chapter Eight: People**
Personal and social responsibility

154	**Appendix I: Site Assessment**	
	Unique features of a property	
158	**Appendix II: Farm Plan**	
	Vineyard practices by season	
160	**Appendix III: Erosion Control**	
	Planning for soil retention	
164	**Appendix IV: Soil**	
	The ground we stand on	
172	**Appendix V: Resources**	
	Information on sustainable farming	
174	**Bibliography**	
178	**Index**	

Acknowledgements

BILL GAFFNEY

While I cannot mention each of you by name, I am grateful to more than a hundred people who helped answer questions and gather material for this book. Special thanks to those whose assistance went far beyond the call of duty:

Phill Blake of the Natural Resources Conservation Service, whose expertise and tireless help made this book possible, and whose personal and professional contributions over the last 20 years seem to have made Napa County's stewardship story more worth the telling.

Those individuals from the Napa Sustainable Winegrowing Group (NSWG) who were exceptionally patient and helpful, and those whose knowledge and critical comments greatly improved the text—especially Kirk Grace, Mark West, DeWitt Garlock, Jason Kesner, Robert Bugg, Richard Camera, Dave Whitmer, Astrid Bock-Foster and Zach

Vineyards in the Watershed

Berkowitz. Also NSWG members Mitchell Klug, for knowledge and insights that cut to the heart of things, and David "Birdman of Carneros" Graves, for multi-faceted savvy.

The entire staff of the Napa County Resource Conservation District, for always finding who and what I needed; Vicky Kemmerer, John Coolidge and Dave Whitmer of the Napa County Agricultural Commissioner's Office, for collecting data and helping it make sense; Shari Gardner of Friends of the Napa River for moral and technical support; Max Seabaugh, for donating illustrations; Tony Bogar and Laura Lamar, for high standards and reliable wit; and lastly, for extraordinary moral support and child care, Lorraine Linstrom, Sarah Di Pietrantonio and Timothy Locke—I couldn't have done it without you.

Community Support

Listed below are the individuals, environmental groups, businesses and government agencies that provided support for this book. The project would not have been possible without these donations, especially the generosity of the Napa Valley Vintners Association, Saintsbury Winery, Robert Mondavi Winery and the Napa County Agricultural Commissioner's Office.

Audubon Society, Napa-Solano chapter

Robert, Jr. and Mary Brown

Cardinale, LLC

Domain Chandon

Friends of the Napa River

Garvey Family Vineyards

Gordon Ranch

Julie A. Johnson-Williams

Britton Tree Services, Inc.

Buckland Vineyard Management

Cain Vineyard & Winery

Casa Nuestra Winery & Vineyards

Clos du Val Wine Co., Ltd.

Etude Wines, Inc.

Franciscan Estate Winery

Kathryn J. & Rainer Hoenicke

Kathy & Tom Meadowcroft

Napa County

Napa County Farm Bureau

Napa County Resource Conservation District

Napa Valley Grape Growers Association

Premiere Viticultural Services

Silverado Farming Company

The Doctors' Company

The Winegrowers of Napa County

UCC Vineyard Fund

Walsh Vineyards Management, Inc.

Lokoya Vineyards

Peter Mennon

John Monhoff

Napa County Agricultural Commissioner

Napa Valley Vintners Association

Nord Coast Vineyard Service

Piña Vineyard Management

Robert Mondavi Winery

Saintsbury Winery

Sierra Club, local chapter

Spottswoode Vineyard and Winery

St. Helena Star

John Thoreen

USDA NRCS Environmental Quality Incentives Program

Weekly Calistogan

Preface

This book is about wine agriculture in Napa County, about managing land with an eye toward the next generation, the next 100 years. But it is most of all a story about people learning how to do things differently. As such, the story neither glorifies nor condemns; it sketches an imperfect work in progress, carried out by a growing number of individuals learning to re-think the ways they manage the land.

Sustainable winegrowing is a complex undertaking driven by the concept that to farm in this Valley—and in these times—means doing a lot more than growing grapes. It means social consciousness and environmental thinking.

For many growers this is not such a big conceptual leap. But those who farm winegrapes are part of an industry with certain conventions. It takes courage (and financial strategizing) to experiment with a more environmentally responsible farming technique. Some growers have to answer to people in accounting offices for the costs of stewardship practices that don't directly increase profits but do improve the non-farmed property areas. Others have to break with family farming tradition. Still others have to part with long-held beliefs about business and agriculture. What they all seem to have in common is a willingness to try and learn something new.

Some of those who appear in this book are beginners and many have been at this for decades. Sustainable farming is drawing all kinds of people for all kinds of reasons—from those who view sustainability as a moral imperative to those who see it as a marketing advantage. Other people see the regulatory writing on the wall and want to prevent further restrictions on agriculture.

Sustainable Winegrowing in Napa County

ERIC GRIGSBY

Regardless of the motives, people are viewing the watershed with new eyes and greater awareness. This new perspective is a significant improvement all on its own.

A few years ago on a newly purchased vineyard property, workers discovered a mysterious trash problem. After picking oilcans, diapers and other waste out of the creek each time it rained, they traced the garbage up the creek to its source. For 50 years under previous ownership, the tributary had been used for household garbage, tires and appliances. After the appropriate permits were obtained, half a century of garbage was removed from the creek. The owner paid $75,000 for the cleanup. It never made the papers, yet the owner was in fact doing a community service while paying for mistakes made by someone before him.

We are all paying. In the past, the Napa River was rife with sewage. For over a century, you could legally dump anything into it. Fish kills were tolerated, and wine slops and city sewage went directly into the river.[1] The stench of rotting fish in Napa once got strong enough to "peel paint from the buildings" along the waterfront, and locals swam in water that carried effluent, factory waste and heavy metals.[2]

We have come a long way in the wisdom of land management, but seemingly innocuous daily activities continuously degrade the watershed. While scientists attempt to measure sedimentation caused in part by development and hillside vineyards, storm drain pollution remains

JULIANE POIRIER LOCKE

illegal but not regulated. Our homes and businesses send everything from car wash products, garden pesticides, lawn fertilizers and laundry soap into storm drains—to the detriment of fish.[3]

Our ever-growing population takes more and more water from the river to run households and businesses, and to water lawns and vineyards. Fish are swimming in lower flows, so when our wastes enter the river via storm drains, a product doesn't get as diluted as it might have when there was more water in the river. The effects on the fish, then, are worse.

Storm drain pollution, together with pesticide-heavy farming practices and poor land management slowly degrade the watershed and all life within it.

Meanwhile we still have fish in the river basin, forest lands in the hills and hope that a growing awareness of shared responsibility for the problems can lead to cooperative solutions, both in the urban and rural landscapes. Some individuals in watershed stewardship groups, environmental groups and progressive winegrowing operations are choosing not to point fingers but to voluntarily assume responsibility and get on with the work of restoring and sustaining the natural resources that grace this county.

JULIANE POIRIER LOCKE

—*Juliane Poirier Locke*
CALISTOGA
DECEMBER 2001

[1] Juliane Poirier Locke, "The Napa River: A sewer ran through it." *St. Helena Star,* March 30, 2000: p. 1.

[2] *Ibid.*

[3] Juliane Poirier Locke, "Household waste contributes to fish kills." *St. Helena Star,* April 27, 2000: p. 1.

Introduction

▲ Undeveloped land preserved on the Valley floor.

Vineyards in the Watershed

The Watershed
Agriculture as stewardship

If the land-use struggles of Napa County—or planet Earth, for that matter—could be summed up in a classic understatement, it might be the headline of a press release issued a few years ago: *"People, agriculture, ecology compete for limited resources."*[1]

The world is getting smaller. The shortage everywhere of unspoiled land and water make land-use issues highly controversial and emotional. Conflict between those who wish to develop and those who wish to conserve natural resources has become so charged that it has been recently described as "the primary tension in human affairs,"[2] greater even than the age-old tension between conservative and liberal.[3]

Introduction

Introduction *The Watershed*

▲ Watershed farming includes stewardship of non-farmed land.

While development and conservation conflicts can separate people, in the same way that politics, property lines, religious belief and social class can drive people apart, it may be a significant coincidence that the land and water themselves are uniting people whose efforts are focused on stewardship of the watershed.

It is within the systems of the watershed that a Napa County farmer learns how to grow winegrapes in a way that degrades "neither the land nor the people."[4] The most important concept to understand about sustainable agriculture is that farming takes place in an area much greater than legal property lines—winegrowers farm in a watershed.

Leaving the comfort zone

The watershed focus represents a shift toward systems thinking. Watershed-centered farming forces a grower out of the comfort zone of convention and into the risk-taking mode of experimentation. And it expands grower responsibilities, so that vineyards are only a part of the whole picture.

Trial and error is the only way to learn what can be accomplished at a site, since every property has unique soils, slopes, vegetation and other characteristics. There is no "recipe book" for sustainable farming and no single right way to do it. But a guiding principle is that farmers are stewards of the land—that includes the non-farmed portions and the land downstream of the legal property lines.

At some properties, the non-farmed areas are wooded or heavily vegetated. At others, the land has been stripped of nearly all of its natural vegetation. Stewardship calls for aiding nature in recovering as much diversity as possible. In some cases vegetation has been removed under current ownership, but in other cases land purchased recently was altered long ago.

Large wildfires, logging, grazing and land cleared by settlers changed much of the watershed before vineyards became popular. Intensive cattle grazing on the Valley floor began in the early 1800s. In an ecological history study of the Napa River watershed[5], biologist Shari Gardner has found records that describe the Napa Valley covered with mature oaks but devoid of underbrush perhaps as far back as 150 years ago.

Triage in the field

Present landowners cannot entirely make up for the land management decisions that were made centuries ago by Native Americans or even decades ago by developers, but they can help stabilize and restore lands critical to the water supply and to the health of downstream lands and users. Every property affects in some way how and to what degree the watershed can support the community at large.

An immediately obvious help to the watershed is to minimize or avoid altogether the removal of trees. Tree removal in the hillside areas is a matter of serious concern, since hillside soils in general are less stable than Valley soils. The remaining woodlands in Napa County represent some of the last vestiges of intact habitats in the watershed.

Woodland habitats not only provide biodiversity and support countless biological functions[6], they act as a recharge zone for the watershed and are thus critical for sustaining groundwater resources. Groundwater serves parts of the rural community as drinking water and for other uses, but it also feeds creeks in some places, discharging during the summer months and supplying water for the stream—and all that live in it—when flow might otherwise dry up altogether. By recharging groundwater, woodlands help sustain a healthy, balanced hydrologic cycle—needed to sustain stream flow and reduce downstream flooding risks.

▲ Homesteader Sarah Wilson, shown here in 1918, was among the settlers who began changing the watershed before vineyards became popular.

▼ Creek vegetation and riparian trees were retained in this vineyard development.

Introduction

Introduction *The Watershed*

▲ Over the Silverado Trail looking southeast toward the Rector Creek watershed in the distant hills. Hills in foreground host unnamed creeks that feed the Napa River. The structure is ZD Winery.

You are here

Your watershed address is defined more or less by slope and elevation. In simplest terms, wherever you are, you're in a watershed—inside a landscape basin (or "bowl") that catches and consolidates rainwater. All the land through which water drains toward the Napa River is part of the Napa River watershed, an area of 426 square miles that contains most of the people and development in the county, including the only cities: Calistoga, St. Helena, Yountville, Napa and American Canyon.

In this watershed, any water that isn't used, stored, absorbed or evaporated on the way travels through the basin into the Napa River, then into the San Pablo Bay, from which it flows to the San Francisco Bay then into the Pacific Ocean.

Most often, water making its way toward the Napa River travels through one of the many creeks, named and unnamed, and through seasonal channels that only transport water during the wet season.

Tributaries often comprise sub-watersheds. For instance, Conn Creek is its own watershed, but is also a sub-watershed of the larger Napa River watershed *(see map, opposite)*. Each sub-watershed has its own unique characteristics, conditions and issues. And each has a different allotment of available water. ■

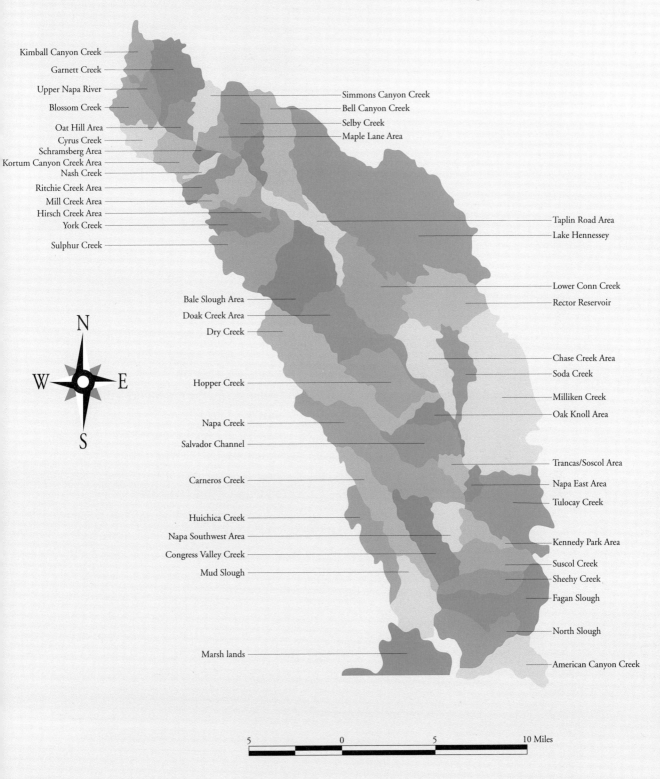

Introduction *The Watershed*

Water speeding up

How water travels within each watershed is key to the stability of people, plant and animal communities. In a well-managed watershed, replete with vegetation and site-specific erosion control, slow-moving water meanders.

At poorly managed sites in the watershed, rainwater moves too quickly—bypassing natural soil storage zones—and flooding is endemic. Fast-moving water damages wildlife habitat, crops and both natural and human-made structures.

Unfortunately, water in the Napa River watershed is moving a little more quickly than it used to. During the last 40 years, the entire main stem of the Napa River has become deeper, cutting 6 to 8 feet downward in places. Ecologically speaking, a deeper river is a degrading river. The process of deepening, or incision, coupled with the loss of secondary flow channels, has put increased pressure on the river to carry higher flows and erode even more.

Development has caused much of this change. Places where the river used to naturally overflow are paved or farmed. When land is pristine, water flows among channels and uneven land surfaces or soaks into the ground becoming temporarily detained on its journey. This process slows the flow and helps prevent flooding. As we introduce structures, dikes, levees, roads and vineyards, these channels and surfaces get lost—bypassed, paved or graded—leaving fewer natural routes for the water.

Cleared lands may also allow less rainfall to be captured on the land, leading to more rapid runoff and to stream flow that is of high intensity and short duration.

The more water that travels and the faster it goes, the more damage that occurs, including floods. Whereas under natural conditions rain might hit the ground and immediately be absorbed by the soil, under urbanized conditions rain might hit the asphalt and immediately pick up volume and speed, traveling toward the river basin.

Preserving what remains

Slowing the flow of water in Napa County — in all the watersheds and sub-watersheds within it—is an effort in which every property and every property manager has a role. To retain trees and woodlands, to avoid creating more impervious surfaces, to leave as much absorbent land as possible and to plant vegetation and trees to stabilize creeks are all positive steps toward preventing further deterioration of the watershed.

Under a sustainable management program, there's a greater chance that the beauty, wildlife and integrity of Napa County will be protected for generations to come. ■

Don't just ignore it—enhance it

All land should be managed. A landowner who sets a few cattle out to graze on non-farmed acreage should not consider the land adequately managed.

Stewardship calls for enhancing the land and helping Mother Nature along. To accomplish this, a land manager should assess the biological resources on the property and learn how a particular site functions in the watershed so a plan can be devised. Areas of undeveloped land may need erosion control, removal of non-native species or planting of native grasses. Wooded areas may need thinning to discourage a hedge of blackberries where a bay tree is trying to grow. Oaks need help reseeding, and the seedlings need protection from grazing or competing plants until the young oaks are well established.

Many landowners have replanted and restored creeks, whether to control Pierce's disease or to prevent bank failure threatening to cut away valuable vineyard land. Under the stewardship model, creek restoration is part of the enhancement required to manage a sustainable property; the more stable a creek, the less erosion will affect the downstream neighbors—both human and fish varieties. Healthy, stable riparian areas maintain water quality and support wildlife species that can aid a farmer in pest control.

Plenty of room for the waterways

The most healthy and diverse of riparian corridors on the Valley floodplain are wide. In the hills, a wide riparian corridor is also better, but the streams occupy a narrower floodplain. It may be that to create a more sustainable property a landowner will choose to maintain a zone of undisturbed land between the vineyard and the creek. Such a zone is commonly referred to as a setback or a riparian buffer.

▲ Because California native oaks are not re-establishing well, some landowners are hand-planting acorns and protecting young oak trees.

Introduction *The Watershed*

Obviously vineyard business profit can only be made on farmed land, but those vines planted close to the immediate stream flood plain can often yield lesser quality fruit or be subject to flood damage, higher maintenance costs and Pierce's disease. With broader setbacks, as part of a watershed-centered management plan, a grower must reassess planting a vineyard too close to the creek, weighing the short-term profits against long-term gains of a healthy waterway on a property that can still be farmed in a hundred years.

Stewardship groups have set the practical example of protecting their watersheds for the future. Pooling their knowledge and obtaining

resources to expand their knowledge of a sub-watershed, stewardship groups are monitoring the erosion rates and stream flows in order to preserve what they have and plan for the future—when there will be more, not fewer, water demands from a growing population *(see Chapter 8).*

Sustainability and terroir[7]

Since the 1990s, wine has taken on a terroir-based identity,[8] which helps spotlight the soil itself. When a terroir-based wine guide advises wine connoisseurs to observe the change in soil types as they drive through the Valley and to "look for changes in color and texture,"[9] a sustainable farmer might take the lesson a bit further and advise getting out of the car to compare the soil health of adjacent vineyards. Soil quality can differ greatly from one vineyard to the next, as influenced by management factors alone.

As sustainability becomes more important to growers and the public alike, it is possible that wine might become identified by an even narrower mark of quality within terroir— that of soil health. It's a reasonable assumption that in the long run a chemically "burned" soil within one of Napa's famous appellations may not produce the quality of grapes grown on healthy, sustainably managed soils in the same appellation.

Given that image marketing based on scarcity and quality increases the price of wine, then the scarce, high-quality wines made from grapes grown in the most sustainable vineyards—in the healthiest soils—ought to be valuable wines indeed.

Learning from experience

Years ago, a portion of this unnamed stream in Carneros was channelized to maximize plantable land and became prone to flooding.

Those managing the site later learned that leaving the stream alone was a better idea. The result was less planted land but more biodiversity, more natural vegetation, less flooding and a more beautiful piece of land. ■

▶ Top: a channelized section of an unnamed creek. Bottom: another section of the same creek, left free to meander. The site is home to numerous bird species.

CHAPTER NOTES

[1] From a press release issued April 26, 2000, by the office of Pam Kan-Rice, University of California Cooperative Extension, Agriculture and Natural Resources Department.

[2] Thomas Berry, "The New Political Alignment," in *The Great Work: Our Way Into the Future.* Bell Tower, 1999: p. 107.

[3] *Ibid.*

[4] John Ikerd, "21st Century Agriculture: The end of the American farm or the new American farm." A paper presented at the Partners for Sustaining California Agriculture, Woodland, March 27–28, 2001: p. 11.

[5] The ongoing Napa Watershed Historical Ecology Project, funded by local private grants, is being conducted by the San Francisco Estuary Institute and the Friends of the Napa River.

[6] One living tree provides $196,250 worth of ecological benefits, whereas a harvested tree has a timber value of only $590. This estimate taken from *Living in the Environment*, 9th ed. by G. Tyler Miller, Jr. Wadsworth Publishing Co., 1996.

[7] *Terroir* is a French term that refers generally to the way in which the soil and climate of a vineyard affect the flavor of the wine.

[8] Jim Lapsley, "Cleansing the Palate: A brief review of California wine history," presented at *Finding the Right Blend: Land Use Planning, Environmental Regulation and the Wine Industry,* Healdsburg, May 23-24, 2001. Land Use and Natural Resources, UC Davis Extension.

[9] Rod Smith, "Terrain and Terroir: A Napa Valley Primer." St. Helena: The Napa Valley Vintners Association, 1997: p. 4.

PHOTO CREDITS
Juliane Poirier Locke: pages 20–21, 22, 23 *bottom*, 27, 28, 29
Collection of de Leuze Family Vineyards: page 24
Collection of Tom Wilson: page 23 *top*

MAP CREDITS
Mapping by Bob Zlomke, Napa County RCD; Jon Lander, Napa County Flood Control District; and Phill Blake, USDA Natural Resources Conservation Service: page 25
Map illustration by Max Seabaugh: page 25

Chapter One

▲ Vineyard erosion delivers
sediment to a stream in
the Mayacamas Mountains of
Napa County.

Vineyards in the Watershed

Erosion

Curbing losses of soil, steelhead and public confidence

Napa County farmers are the most highly regulated winegrowers in California[1] in part because of the hillside ordinance that growers accepted—some more willingly than others—as a necessary safeguard.

For some growers the added time and expense of regulating hillside erosion control made sense both for the watershed and for the industry as a whole. In effect, the conservation regulations adopted in 1991 established an industry standard that would protect both the hillsides and the community from any number of mishaps, caused by poor management practices and heavy rains, that would seriously erode not only the landscape but the confidence of the public.

Chapter One *Erosion*

▶ Concentrated flow erosion occurred in Wild Horse Valley, Napa County, in 1983.

Two such mishaps, which occurred during winter storms in the northern Napa Valley, catalyzed support for a hillside ordinance.

Erosion takes to the highway

In 1980, heavy rains on a newly deforested hillside drove a slug of silt down to the Valley floor. When the silt hit Highway 29 near Calistoga, it became obvious to the public that there was a problem in the hills. County supervisors appointed a blue-ribbon panel to look into the erosion issue, but hillside development went on as usual for another nine years—with a few exceptions.

The exceptions were those who voluntarily began implementing erosion controls following a soil-loss study published in 1985 by the USDA Soil Conservation Service.

The SCS (now the Natural Resources Conservation Service), after realizing that few were heeding its advice, had given up trying to stop hillside vineyards and had begun instead to develop ways to control erosion. The agency's co-authored regional study of soil erosion in both Sonoma and Napa counties showed annual erosion rates as high as 350 tons per acre.[2]

While some growers and consultants began voluntarily adopting erosion control practices in their hillside vineyards, others did not. After another vineyard development washed downhill

▲ This sheet flow erosion is picking up sediment in suspension.

Continued on page 36

Case Study: Controlling hillside erosion
Napa River/York Creek watershed

▲ A century-old fig tree is preserved in this vineyard block.

◀ A vine row drop-inlet catches storm runoff on hillsides at Spring Mountain Vineyards.

Winegrowing on the slopes costs more per acre and poses more threat of erosion than flatland viticulture. Sustaining hillside farming in Napa County includes the cost of installing and maintaining erosion control systems and the practice of leaving areas of woodland vegetation intact. In this case study, vineyards make up only 26 percent of the total property.

On a dry spring day, Rex Geitner points to the ground where storm drains the size of large pies rest in a pattern across the slope. Beneath this hillside vineyard lies an erosion control labyrinth—miles of underground drainage pipe designed to catch and slow the speed of storm water before it can carry off soil. Above the ground, no-till cover crops help keep soil rooted in place.

These soil conservation practices are somewhat new to the site. When Geitner started managing Spring Mountain Vineyards in 1994, he took over a property with stunning views, the elegant architecture of a bygone era and erosion problems that date back more than a century.

For over 130 years winegrapes have been grown on the site, in ways that previous farmers believed to be the best practices. The clean cultivation, or disking, of hillside vineyards—including terraced vineyard blocks on slopes exceeding 48 percent—was standard practice, and over the years it contributed to soil loss, sediment pollution, terrace failure and stream bank collapse.

Geitner had his work cut out for him. He doesn't blame the erosion problems on those who farmed there before him. Geitner claims he's made his own share of mistakes in the 28

▼ High above St. Helena, Rex Geitner stands in non-terraced, east-west running rows of Cabernet Sauvignon.

Chapter One: Erosion

▲ An underground pipe, four feet in diameter, discharges hillside runoff into a settling basin fitted with sediment traps and armored flow lane.

▶ An armored waterway with drop-inlet captures runoff beside a vineyard road.

years he has farmed hillsides, and that trial and error are one way growers learn to improve their practices. In recent years, social forces have also changed farming practices.

One community-driven force that changed hillside farming practices in Napa County was the adoption in 1991 of conservation regulations, which mandate hillside erosion controls.

Farming according to the conservation regulations was "painful at first," Geitner recalled, because the new way "wasn't the way we were used to doing it." But he now concedes that the regulations have not only given rise to a new conservation technology but have "made us all better farmers."

County-regulated erosion control "brought about practices we might have dreamed about but not implemented," Geitner said. "It brought a greater awareness of a need for preserving what keeps us in business—the soil and its surrounding environment."

Like other hillside winegrowers in Napa County, Geitner has to manage both for winegrape quality and watershed integrity. On this 850-acre property—formerly four separate Spring Mountain estates—the primary watershed goal is preventing sediment from reaching York Creek via any of the three tributaries.

To manage erosion and water quality on the eight-parcel site, Geitner in 1996 took steps to create and implement a master erosion control plan for the 225 hillside vineyard acres. "I knew it would be a huge undertaking," he said, "and a very expensive one."

He was right on both counts. The erosion control systems—for which he credits Phill Blake of the NRCS, Dave Steiner of the Napa County RCD and Silvia Toth, a former county planner—are part of an erosion control

project Geitner describes as "large, complicated and very do-able." The project cost millions of dollars.

Spring Mountain Vineyards is now participating in an erosion control study to measure vineyard runoff from cultivated slopes where erosion controls have been completed.

At one vineyard block, 15 acres of non-terraced vineyard drain into a single four-foot culvert pipe. Runoff shoots out of the pipe and hits a set of fieldstone stairs embedded in a seasonal channel. The stairs slow the water and let it drop its sediment load before flowing into York Creek. At a terraced vineyard, rainwater is caught by culvert piping and carried to a spreader that redistributes the water to a hillside

where it can be released slowly and absorbed by the ground.

Geitner feels fortunate to work with a landowner who supports good management practices and will pay for

them. "I can imagine others thinking I'm up here playing in a big sandbox with unlimited funds," said Geitner. "I am in a big sandbox, but the funds are not unlimited. People need to understand, however, that you don't go into this business on a shoestring."

Geitner routinely budgets $20,000 per acre for erosion control—a third of the per-acre cost of replanting at this site, where only 26 percent of the property has been converted to vineyards and the rest provides unfenced wildlife corridors. Geitner believes that with the right management practices, winegrowing can be sustained at this site for centuries to come.

"I have been to other parts of the world, to the steepest sites, where they do what it takes to preserve the soil," said Geitner. "I've seen where the same hillside site has been farmed by the same family for over 500 years—that's a good start at sustainable farming." ■

▲ Failed terrace vineyards in background contrast with non-terraced replants in foreground. The new vineyard has erosion controls and an eight-fold vine density increase—from 500 to 4,000 vines per acre.

◀ T-spreader with rock apron reduces the speed of collected runoff, redistributing the flow into smaller, slower rivulets that can be absorbed by the adjacent woodland areas.

Chapter One *Erosion*

▲ From cleared vineyard land above Bell Canyon Reservoir, almost 2,000 tons of sediment slid into the lake.

Continued from page 32

and grabbed public attention, hillside stabilization would no longer be a voluntary practice.

The public sees red

During a 1989 October rainstorm in the mountains northwest of Angwin, erosion from 30 acres of cleared hillside flowed directly into a city reservoir.

Russ Walsh, chief water plant operator at the Bell Canyon Reservoir, was on duty at the time of the disaster and recalls how the eroded vineyard soil appeared in the water supply for more than 5,000 St. Helena residents.

The soil had carried a telltale pigment into the water, lending an eerie tinge to the crisis. "Nobody knows how much silt went in," said Walsh. "All I know is that we had red water."[3]

The erosion caused severe turbidity and triggered a public emergency and a chain of reactions including responses from residents, the media, natural resource agencies, city government and the district attorney. Almost 2,000 tons of sediment had entered the lake.[4]

Erosion from hillside development was, evidently, a threat to public safety. Something had to be done.

Phill Blake, the county's NRCS district conservationist, drew up emergency erosion control plans for the two offending vineyard properties, effectively controlling the sheet and rill erosion that had come down the mountainside. Those plans would later become templates for the hillside ordinance.

County supervisors asked winegrowers in 1990 to collaborate with county, state and federal agencies to forge conservation regulations. After extensive hearings, the document created by the task force was adopted the following year. The new ordinance mandated that developing a slope

Continued on page 38

Sheet erosion: the invisible foe

Even without the obvious scars of deep-cut gully erosion, every property in the watershed is susceptible to soil loss, particularly to the insidious, almost invisible type known as sheet erosion.

A certain amount of "background" erosion, anywhere from one to five tons per acre annually, is natural, according to Dave Steiner, soil conservationist with the Napa County RCD. Steiner says the amount of soil leaving a site in the rain is influenced by the amount of organic matter in the soil and the amount of vegetation growing on the surface. Whether that surface is sloped, and to what degree, is also an important influence.

"At most sites in Napa County, if you're losing seven tons per acre each year, your erosion rate is not sustainable—and there's a good chance your site is creating water quality problems downstream," Steiner said, adding that some people might lose five to seven tons of soil per acre each year and not be aware of it.

"Sheet erosion is barely perceptible, if at all," said Steiner. "The energy of flowing water displaces and moves very small amounts of soil particles. It might move them an inch, or it might take them to Mare Island. It's all a matter of energy."

Sheet erosion peels one ton of soil off an acre of land in a layer no thicker than a few sheets of writing paper. Over a generation, the topsoil on a property could erode away completely without anyone noticing.

But noticing—and reducing—erosion, in all its forms, is a basic tenet of sustainable farming and land stewardship.

The RCD and the NRCS in Napa provide a free service to anyone interested in assessing the net soil loss on a particular site. ■

▲ 1. This gully erosion occurred in Wild Horse Valley, Napa County, in 1983. In a terraced vineyard with no erosion controls, gullies formed within 72 hours during winter rains.

2. Underground streams flowing from Mt. George set up conditions for this 11-acre landslide in Coombsville, near Mt. George, in 1983. A sinkhole at the upper end of the vineyard was over 20 feet deep.

3. At this unidentified location in Napa County, sheet and rill erosion in 1984 caused erosion rates equivalent to 15–25 tons per acre.

Chapter One *Erosion*

▲ This vineyard clearing was done in the Howell Mountain area.

Continued from page 36

of 5 percent or more required a permit; a permit required a county-approved erosion control plan.

As a result, the sediment pollution to the Napa River watershed went down and the cost of hillside farming went up.

Hidden costs of hillside development

Farming the hillsides is expensive. While developing an acre of Valley floor may cost $32,000, an acre of hillside vineyard can cost up to $80,000 to develop.[5] That cost estimate factors in erosion controls required by the county.

Dave Steiner, a winegrower for 30 years who evaluates Napa County erosion control plans, believes that the extra cost of hillside erosion control is not too much to ask. "If we can't afford to do it right," Steiner said, "we shouldn't do it."

Steiner, who has walked hundreds of vineyard sites in Napa County, observes that if a developer fails to make the necessary investment in hillside stabilization, "he's asking his downstream neighbors to carry the cost of his development."

The downstream costs of eroding hillside vineyards are more than monetary. Federal, state and local agencies, as well as regional and local citizens' groups, are now conducting watershed studies that may affect agriculture.

The groups involved in these studies share a concern for the habitat of migratory fish, including threatened steelhead trout. Among other things being investigated is to what extent the sediment pollution in the Napa River watershed is harming fish habitat.

Sediment typically affects the home territory of fish by filling up spaces in the gravel bed, smothering the fish eggs. It can also change the water quality—the temperature, visibility and oxygen content—enough to harm or kill fish.

While the quantities and effects of sediment in the watershed are being studied, the hillside regulations continue to require a plan to mitigate erosion. Erosion control plans are approved if the plan will keep the soil on the hill, Steiner explained. What county approval does not do, according to Steiner, is address or calculate for "the lost value of wildlife or the ruined viewshed."

How much is enough?

Concern for wildlife and for the natural state of the Napa County hills is a factor in future vineyard development, in terms of land ethics, community relations and regulation.

While the public observes new hillside sites being deforested, the most commonly expressed fear is that all the hillsides will one day be vineyards. Members of the local Sierra Club have voiced this fear at public meetings and in their literature.

Seeking a moratorium on hillside development, the Sierra Club sued the county of Napa in 1999 on the grounds that the hillside ordinance skirted the California Environmental Quality Act (CEQA).

The lawsuit was settled in favor of the Sierra Club; consequently, as the county evaluates applications for hillside developments, it must provide a much greater degree of environmental scrutiny than was previously required.

While the debate continues over how much more development the area can sustain, erosion in Napa County has been reduced in the past decade through education, legislation and a growing consciousness about watershed processes.

Trusted experts

Although the RCD and NRCS are public agencies, they do not act as soil police. "We do not turn in violators," said Phill Blake, NRCS conservationist. "That's up to the public. We're here to inform and inspire, not to require."

Staff specialists from the two resource agencies help property owners and managers look at the way a site functions in the watershed, including the runoff patterns —what they were under natural conditions and how they might be affected by current and proposed management practices.

The agencies offer site-specific erosion analysis at vineyards, presenting alternative practices and sometimes cost-sharing. Ranchers can learn about practices that will protect streams from the impact of grazing.

"For the most part, people who know us well trust us," said Blake, who estimated that half the acreage in Napa County, including both ranches and farms, has had some kind of input from either the RCD or NRCS. "Most of what we do is strictly educational," Blake added, "and one-on-one assistance to help people to understand watershed processes and to learn how they can become better stewards of the land." ∎

Chapter One *Erosion*

▲ 1920s photo of a hillside vineyard facing east from Spring Mountain toward St. Helena. This photo shows vineyards owned by the Sheehan family, planted before Prohibition. The property still sustains winegrapes although some of the vineyards have been reclaimed by nature.

CHAPTER NOTES

[1] Janet C. Broome, et al. *Grower's Guide to Environmental Regulations & Vineyard Development,* Sustainable Agriculture Research and Education Program, 2000.

[2] "Hillside Vineyards Unit Redwood Empire Target Area: Napa and Sonoma Counties, California." United States Department of Agriculture Soil Conservation Service River Basin Planning Staff, 1985.

[3] Russ Walsh, chief water plant operator at Bell Canyon Reservoir, interview with author, St. Helena, March 7, 2001.

[4] The Soil Conservation Service estimate, calculated in cubic yards (1,400), converts to 1,900 tons.

[5] Mike Fisher, "High Intensity Grape Growing," in *Vineyard Economics,* Motto, Kryla and Fisher, 2000: pg. 27.

PHOTO CREDITS
Phill Blake: pages 30–31, 32, 37 *middle, bottom*
Chip Bouril: page 36
Juliane Poirier Locke: pages 33, 34, 35, 41
NRCS: page 37 *top*, 38
Collection of Tom Wilson: page 40

▲ Cover crops, required in all Napa County hillside vineyards, help control erosion and improve soil health.

Chapter Two

▲ A pool in the Napa River, late summer.

Water
Conserving a finite supply

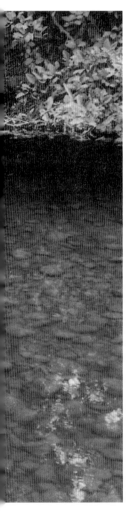

As clean water becomes scarcer and more expensive, and as groundwater tables continue to drop or become contaminated, there is growing concern about how finite water resources might be managed efficiently and fairly.

Most farmers and rural residents in Napa County, unlike urban water users, rely on surface water, groundwater and small, privately owned reservoirs of stored rainfall. While cities maintain large reservoirs and negotiate contracts for imported surface water, the water feeding rural needs is a more complex and tenuous supply—vulnerable to the whims of nature, the complexities of geology and the obscurity of the water rights system. Most significantly, rural water supply is limited and depends on rainfall for replenishment.[1]

Chapter Two *Water*

▶ A healthy section of the Napa River. Flows and water levels are affected by development in the watershed.

Where the rain goes

Average rainfall in Napa County is from about 55 inches a year in the hills north of Calistoga to just over 20 inches in Carneros. As that water makes its way through the various kinds of rock formations, it is almost impossible in some areas to know whether two close neighbors are getting water from the same source—underground flows can go in all directions.

Like rainfall, water supply varies at sub-watersheds throughout the county. Thus, a sustainable water management plan for both agricultural and residential sites depends on knowing what's in the water budget and how to stay within that account.

Overdrawn water accounts

When water is not budgeted, homes and vineyards are at risk of running short. Already in some areas of Napa County people are using more water than their sub-basins are getting from local rainfall. In some rural neighborhoods, tributaries have been pumped dry on occasion—killing fish and depriving some users of a fair share of water.[2]

On one tributary in southwestern Napa County, when an over-draft reduced flows and killed fish, investigators found that illegal water pumps had been installed by homeowners who were extracting stream water to irrigate their lawns.

▶ Robert Adams displays a trout catch from York Creek, a tributary of the Napa River, in 1959.

Vineyards in the Watershed

In another sub-basin, groundwater supplies are deficient to the point that residents must restrict water use to a very modest level. In the area of Mt. Veeder where springs provide the water, residents reportedly truck in potable water at a cost of $165 per 3,300 gallons. A vineyard in the same area, which has increased its water needs beyond its water budget, has been trucking in recycled water for the past few years in order to irrigate.

In other sub-basins such as Carneros Creek, neighbors—both residents and growers—are taking proactive measures. They are tracking the water flows in their tributaries and adjusting their water use to conserve supply and help protect fish.

How much flow?

Public agencies and interest groups are trying to determine just what kinds of flows are needed for fish life, particularly steelhead.

A study conducted by the Friends of the Napa River found that steelhead trout populations exist in more than half of the tributaries distributed throughout the watershed.[3] Because steelhead trout are listed as an endangered species, protecting their habitat may restrict future water use.

The water study that may have the most impact on Napa County life is a federally mandated study being conducted by the Regional Water Quality Control Board.

The four-year study is attempting to identify and rank the problems of the watershed, ultimately determining what the watershed can bear each day—the total maximum daily load (TMDL) from each impairing source.

The local TMDL study, mandated by the federal Clean Water Act, will update a study conducted in the 1980s. The earlier study determined that the Napa River watershed was impaired by sedimentation, nutrients and pathogens; the current study will determine to what degree those conclusions are valid. The current TMDL will also look at other factors affecting

▲ The Napa River watershed provides habitat for the steelhead trout, listed by the federal government as an endangered species.

Chapter Two: Water

Case study: Neighbors and stewardship
Napa River/Carneros Creek watershed

▲ Blaine Jones of the RCD installs a telemetric stream gauge.

▶ David Graves in Carneros Creek.

The Carneros Creek land stewardship group is using experts and grant money to uncover the history of their creek, restore it and make sure the water supply doesn't get overdrawn. In the process, some members are becoming better neighbors.

"The great thing about the stewardship group," said David Graves of Saintsbury Winery, "is to have the people who live here and those of us who work here getting together and talking." The membership is made up of Carneros Creek residents who don't farm, residents who do farm and "people like me," Graves explained, "who don't live in the watershed but own land and farm in the watershed."

For some of the neighbors the stewardship group has provided the first opportunity they've had to talk to farmers about chemicals and water use. And some of the growers have had their first opportunity to talk to residents about restoring the riparian zone with species that don't host vectors for Pierce's disease.

With help from CALFED[4] grant money and expertise from the Resource Conservation District (RCD), private consultants and the U.S. Geological Survey (USGS), the Carneros Creek stewardship group has set up a voluntary well-monitoring program to assess groundwater resources. "We don't want to end up overdrawn," Graves said.

The stewardship group is investigating physical properties of the

▼ David Graves, David Dowling and Walter Gentry, members of the Carneros Creek stewardship group, monitor a Carneros well.

watershed as well. "We're looking at the creek to understand which reaches have stable banks and which have problems," Graves explained. "We want to map the sedimentation sources, look at the surface water supply and use, and then we are also very interested in historic land uses—the history of grazing for example. We're going to pay for some detailed studies in our watershed."

Meeting at the local elementary school every few months, the loosely organized group has done a lot. "We started in February and we're already rockin' and rollin' here," Graves said. The progress the group has made in such a short time, he observed, "would not have been possible without the enthusiastic participation of the stakeholders."

Graves himself is not short on enthusiasm, both for the process and for the place where his vineyards grow. "Did you know," he asked, "that birders have identified 130 species of birds in the Carneros region? Including golden eagles. I've seen them myself."

Graves recommends stewardship groups to anyone who has concerns about bank stability, water quality or trees. He suggests that anyone who wants to learn how to run a watershed-responsible household, how to be more fish friendly, "even finding what the creek used to look like," should call the RCD and learn how to participate in a stewardship group. In addition to self-education opportunities, Graves says, a stewardship group is valuable for getting neighbors together "just to talk to each other." ■

Those interested in forming a watershed stewardship group may get free assistance by calling the RCD at (707) 252-4188.

"The great thing about the stewardship group is to have the people who live here and those of us who work here getting together and talking."

—David Graves

Chapter Two: Water

Chapter Two *Water*

watershed health as it tests the hypothesis that sediment is impairing fish habitat in the Napa River watershed.[5] Data collected so far has not precisely supported that claim *(see page 50).*

Many people are waiting for regulatory action to follow on the heels of this study in 2004. Others are taking stewardship action now.

Sustainable water budgets

In some places around the county, neighbors are cooperating in land stewardship groups that correspond to the sub-watershed in which they live and work *(see map, page 25).*

The Sulphur Creek stewardship group in St. Helena, for example, is among those developing a sustainable water budget that maintains creek flows adequate for sustaining fish populations.

Each sub-watershed has a different supply of water, determined by many site-specific factors including soils, geology, local rainfall, steepness, rate of rainfall infiltration, water flows and vegetation.

With help from the Resource Conservation District (RCD) and the Natural Resources Conservation Service (NRCS), neighbors are forming stewardship groups that tackle water use budgets. Volunteers in the southern part of Napa County are setting up stream gauges as a means of determining when pumps should be turned off to maintain adequate stream flow for fish. The newly formed Carneros Creek land stewardship group *(see page 46)* is developing a water-use model based on stream flow—as the Huichica Creek land stewardship group did in the early 1990s *(see page 149).*

Dry farming

Historically, vineyards in the Napa Valley were dry farmed. In Europe—including the southern regions where weather patterns are similar to Napa County—vineyards are not irrigated. At some places, such as the Tofanelli Vineyards in Calistoga, vineyards are dry farmed as they have been for generations.

Continued on page 53

▼ Drip irrigation in a vineyard.

▶ Martin Mochizuki of Walsh Vineyard Management.

Tools for deficit irrigation

It's a common misconception that all agriculture is a glutton for water. While many crops in California require flood irrigation, vineyards do not; excess water is bad for the vines. To target water use, a few commonly used technologies help grape growers achieve what is known as deficit irrigation.

Terms like "pressure bomb" and "neutron probe" might conjure images of old James Bond movies. But these two devices ride placidly in the back of Martin Mochizuki's red pickup truck as if they were no more dynamic than a couple of over-sized battery testers.

The ordinary-looking pressure bomb and neutron probe, at a combined cost of almost $8,000, help growers do two things: irrigate in a timely fashion and use only the minimal amount of water on the vines.

Mochizuki, of Walsh Vineyard Management Services, sits at a vineyard picnic table in Oakville, where the Cabernet Sauvignon clusters are still hanging on the vine, within days of harvest.

With both hands Mochizuki sets the weighty pressure bomb on the table and explains what it does. A leaf positioned in full sun is cut and placed inside a plastic bag and put into a pressure chamber with its stem sticking out of the chamber. After pressure is applied, water starts to bubble out of the stem. The tension employed to pull water through the vascular system of the plant is recorded in negative bars—measures of pressure that correlate with the water stress in the plant. Knowing how much water stress the leaf exhibits allows the grower to determine when to irrigate. Without this tool most growers have to guess when to begin irrigating, often starting earlier than necessary.

Turning from the pressure bomb, Mochizuki holds up a device that can read how much moisture is in the ground—the neutron probe. A database containing years of historic data collected from the probe, Mochizuki explains, can help you "know within a couple of weeks when you need to water." The pressure bomb then "helps you fine-tune the start of irrigation."

Mochizuki's company uses the neutron probe and the pressure bomb for irrigation planning on all of the 2,200 acres it manages in Napa and Sonoma counties. It also contracts with owners of 500 more acres just to provide the probe and pressure bomb field data which, combined with UC Davis weather station data, are the basis for irrigation decisions.

Each week that the vines are irrigated vineyard managers replace an average of 40 percent of the water used by the vines the week before. Chardonnay can take the least amount of stress, and reds can take the most. The goal, Mochizuki says, is to "use the least amount of water to get the highest quality of grapes." ■

TMDL in Napa County: *An interview with Mike Napolitano*

▲ Mike Napolitano at a restored reach of the Napa River along Frog's Leap's Galleron Ranch in Rutherford, where owner John Williams hired a restoration specialist to repair erosion damage.

Some stakeholders are worried about the outcome of the TMDL study to be completed in 2004 *(see page 45)*. A common fear is that the requirements of fish and other wildlife will take precedence over residential or agricultural water needs.

"You usually don't get into anything more controversial," said Mike Napolitano, "than telling someone how much water they have or don't have."

Napolitano, of the Regional Water Quality Control Board (RWQCB) in Oakland, leads the TMDL study in Napa County—a process he hopes will help "maintain a vibrant economy while restoring fish habitat so there are self-sustaining populations that aren't in jeopardy."

Although the TMDL study results will not tell anyone how to divvy up the finite water resources in Napa County, it will likely have an impact on future water management as it relates to supporting habitat for fish—especially steelhead trout.

The scientists conducting the local TMDL are testing the hypothesis that sediment is causing significant adverse conditions for fish in the Napa River watershed. Evidence collected so far indicates that not sediment alone but a number of issues are likely to be undermining the viability of fish populations. Water quality degradation may in fact be more closely linked to flow and water temperature than to sedimentation.

"The amount of flow in the dry season, in tributaries in particular, may be significant for restoring steelhead and other species," Napolitano said, adding that trust and participation from landowners will be critical to the success of the study, and he can empathize with landowners who feel uncomfortable cooperating.

For landowners to help a trustee agency can be "a very hard thing, given how threatening this can feel to people," said Napolitano. "It's much harder than anything we've had to do before, working on flow issues. The best job we can do is to understand the issues and the science in order to improve reliability of supply and stream flow."

He would like to see a majority of stakeholders involved in the process, trusting that the researchers are taking a balanced approach, based on good science—which will always contain some uncertainty—and consistent with the law and with the communities. "We're not going to do anything radically conservative or liberal," Napolitano said.

Impediments such as small dams, diversions and road crossings make it difficult or impossible for steelhead or other migratory fish to move up or down at key times, maybe cutting off fish from habitat. If such impedi-

◀ Undisturbed bend in the Napa River, 2001.

If water supply reliability is found to be an issue, public-private partnerships and grant money might be available to help change the methodology and timing of groundwater or surface withdrawal. One hypothetical possibility is to modify water rights so direct diversions could be taken in wet seasons rather than dry; diverted water could be stored for use in dry seasons when water flows are so low that diversions would drastically reduce flows.

The team is getting very close to completing its work in the watershed, and Napolitano feels that remarkable local leadership is needed to get the job finished—leaders getting stakeholders to collaborate. Having observed a high level of stakeholder involvement in the Napa River watershed, Napolitano remains hopeful. "There's been a lot more (people helping) preserve the heritage and economy of the watershed than people pulling each other down," he said.

▲ Locals take a swim in the Napa River near Bale Lane in Calistoga during the early 1940s.

Chapter Two: Water

ments are found to be significant, fish passage improvement projects might be the solution. Such projects could be funded with public money. (The federal Clean Water Act has granted $90,000 to Napa County watershed improvement projects since 1991.)

Ideally, community members would be part of the measurement team. A network of 50 continuous recorders to measure water levels throughout the watershed would help researchers learn how people are managing water. Permission to collect flow data on private property is critical to a thorough study.

"I think the stewardship groups are great," said Napolitano, who thinks there should be more groups and more money for them. "There's no question that the thinking they've come up with is going to be the most important part of any success. We need to have them in all the key tributaries."

Human nature

"What will make or break (the process) is the relationships among people locally," said Napolitano. "We can try to do a lot of things, but the more confidence people have locally, the better chance we have."

Napolitano observed that since nobody likes being told what to do, the best approach to start with is offering incentives to improve the watershed. "It always works better when people have a chance to decide for themselves what's in their best interest. We have a number of carrots, we like to use those first," Napolitano said, adding that in the world we live in, there has to be the threat that people can be regulated or that they can be taken to court. Without that threat, "there will always be a small minority that isn't willing to be responsible," he said. "It's just human nature." ■

▲ A Carneros pond preserved on vineyard land.

Continued from page 48

Although many growers elect to irrigate vineyards for numerous reasons, sustainable farming requires each grower to consider dry farming as a possible alternative to irrigation. As long-time grower Mitchell Klugg points out, irrigation should not be an unquestioned component of a farming plan any more than pesticides should be. "Each grower needs to do the calculation," said Klugg, "to determine which of the two practices is more sustainable, using economic, environmental and qualitative criteria for the decision."

Conserving water

But it is not a choice between dry farming and intensive watering. Since they do not require flood irrigation or sprinklers, grapevines can be grown with minimal irrigation. To use their water supply carefully and to water with precision, many vineyard managers employ soil moisture and leaf stress measurement technology, which aid in efficient water use *(see page 49)*.

Where water is required, it is most commonly applied through a drip hose. When drip irrigation became common in Napa County during the late 1970s, along with new training and trellising methods, poor soil could be farmed successfully and high-density plantings could survive on hillsides with soils only two feet deep. (Planting density increased from 400-600 vines per acre in the late 1980s to over 1,000 vines per acre in the late 1990s.)

To capture and use water that would otherwise leave the property, some growers are retaining water from the shallow groundwater table that fluctuates seasonally but remains fairly high in Yountville, Calistoga and the Oak Knoll area of Napa. In the past, growers drained this water via underground pipes—called drain tiles—away from the roots of the vines to keep them from getting "wet feet." Shallow groundwater that was once drained into ditches and discharged into the river is now increasingly stored in ponds for re-use as irrigation water.

Groundwater conservation is also promoted by healthy soil, particularly during rainfall. Poorly managed soil is frequently compacted and can increase runoff and contribute to erosion, stream sedimentation and flooding events, but properly managed soil allows higher infiltration rates by which water can percolate through the soil into deeper layers, depending on the soil.

▲ In this 1946 photo, St. Helena tomato farmer Dick Myrick rests on a culvert pipe through which river water is being used for flood irrigation, a practice not used in vineyards.

Chapter Two *Water*

> "The amount of water it takes to irrigate one acre of lawn will irrigate 12 acres of winegrapes."
>
> —Tim Healy

Another means of conserving groundwater resources is the use of recycled water.

Recycled water for irrigation

In the water-poor region of Carneros, where the vineyards of Domain Chandon occupy land in both Napa and Sonoma counties, vines have been irrigated with recycled wastewater from Sonoma County for more than eight years.

Domaine Chandon's reservoirs collect surface runoff during high rain, about 150 acre-feet. And about 250 acre-feet of recycled water are directly available to the irrigation system. "We can turn a valve one way to pull from the reservoir and the other way to pull from the recycled water system," said Dana Zaccone, Domain Chandon's irrigation manager.

Zaccone says that by using recycled wastewater Domain Chandon is helping the watershed. "We're allowing the Sonoma County Water Agency to utilize their treated water," he said, "rather than storing it or pumping it into a slough that eventually drains into the San Pablo Bay."

As more growers are added to the system, users may be required to store the recycled water. And when the treated wastewater achieves commodity status, Sonoma County foresees developing a fee structure, but not anytime soon.

In Napa County, recycled wastewater is sold by the cities of Napa, Yountville, Calistoga and St. Helena. American Canyon has plans to sell treated wastewater as well.

In the dry season, the Napa Sanitation District has about 2 million gallons available at 75 cents per 1,000 gallons—about $250 an acre-foot. The charge, according to Tim Healy, an engineer for the district, is to help cover the costs of tertiary treatment and pipeline.

Wastewater can be treated at three different levels, with tertiary being the highest quality. Although secondary treatment is safe for vineyard irrigation, the district takes the water to the highest level, which costs more.

"Vineyards take a lot less water than turf," Healy said. "The amount of water it takes to irrigate one acre of lawn will irrigate 12 acres of winegrapes."

Some experts predict that treated wastewater will soon become a commodity, especially valuable during times of drought, when the watershed is not being replenished by rainfall.

▲ This Chiles Valley reservoir is managed without chemicals and serves as a bird sanctuary.

Looking ahead, acting now

While water use per vineyard acre is relatively modest compared to other types of agriculture, the number of vineyard acres in Napa County most likely cannot continue to grow without a careful look at how much water is available—and sustainable—within each sub-basin. It is expected, for example, that new developments in Napa County will be evaluated for—among other things—available water budgets.

The outcome of the TMDL study may change how water is allocated, yet proactive stewardship in each tributary is likely to influence how the TMDL data are interpreted and applied in the watersheds of Napa County.

Protecting soil and water quality

Heaters

Water quality is protected largely by good management practices. Training workers to prevent spills or leaks of toxic materials including fuels and pesticides is a critical responsibility of growers.

Vineyard return-stack heaters, which are used to protect grapevines from frost, contain diesel fuel, a potential threat to water quality. Fuel spills from these devices can be caused by flooding, damage, upset from wind or vandalism and filling operations. Flooding is the most serious of risks, because it can upturn even properly stored heaters and carry diesel fuel directly into surface water. In 1996, soil tests and a grower survey in Napa County suggested a limited risk of groundwater contamination from heaters.[6] However, the filling, transport, use and storage of heaters do present the risk of limited soil contamination.

Using an alternative form of frost protection is a more sustainable means of preventing environmental contamination from diesel fuel. Viable alternatives to and recommended management practices for return-stack heaters can be found in the Napa River Watershed Owner's Manual, obtained through the Napa County Resource Conservation District (RCD) office.

◀ Return-stack heater

Backflow and spill prevention

The possibility of spills and backflow accidents during mixing and loading of pesticides can present a threat to workers and to soil and water quality. The best site for mixing, loading and storage of hazardous materials (including the storage of fuels and fertilizers) is as far as possible from waterways.

▶ Overhead fill spout with an air gap.

Vineyards in the Watershed

Agricultural plastics recycling

Properly cleaned and recycled pesticide containers help keep toxins out of the landfill and the water supply.

The plastic containers in which pesticides are purchased can pose a threat to soil and groundwater quality when they leak pesticide products into the landfill.

To mitigate this risk, the Napa County Agricultural Commissioner's office launched an industry recycling program in 1994 for the recycling of agricultural pesticide containers, irrigation drip hose and cardboard.

Grower participation in the recycling program more than doubled in six years, and almost 100 tons of plastic have been recycled to make new pesticide containers and other industrial—not household—products.

▲ Almost 100 tons of plastic have been recycled from Napa County agricultural recycling efforts since 1994.

Plan for a worst-case scenario
Plan ahead for accidents such as a tank leak when establishing a mixing, loading or storage site. Also, develop procedures for preventing materials from moving off site into a creek or a well. Mixing and loading on a pad, for example, allows for ease of cleanup when there are spills and protects the soil and the well.

Use an overhead fill with an air gap
This prevents hazardous materials from getting sucked back into the water source if something goes wrong with the water pressure. Some sites have a reverse pressure (RP) device that prevents backflow contamination of groundwater and surface water. Such a device is mandatory where pesticides are injected via irrigation systems.

Maintain a wellhead berm
While this measure is required at every site, it is particularly critical on sites where there are older wells. Newer wells have protective casing, while older wells are more vulnerable to spills. Contaminants that accidentally enter the soil at just 12 inches of depth can move laterally into the groundwater where there is no well casing. Where there is casing on a well, contaminants are less likely to enter the groundwater table. ∎

These recommendations are from the Napa County Department of Agriculture.

Chapter Two: Water

Chapter Two *Water*

Training workers to protect water resources

◀ Daniel Robleyto, of Sonoma Grapevine Nursery, teaches farm workers about pesticide sprayer calibration at a community workshop.

Trained workers who grasp the reasons for water-quality protection are an asset to business, to watershed farming and to the community.

Water-quality protection strategies in Napa County are carried out largely by farm workers in their day-to-day work. Consequently, the education and training of workers is essential to any degree of sustainable agriculture—both for the good of the watershed and the good of the workers.

A trained worker knows why excess pesticide solutions are not poured on the ground, why diesel fuel should never be allowed to leak into the soil and why erosion control is critical in all phases of a farm operation. Knowledgeable workers can act as stewards to protect water quality.

The Napa County Agricultural Commissioner's office, the Napa Sustainable Winegrowing Group (NSWG) and other agricultural groups sponsor worker training workshops, many of which are conducted in Spanish.

▲ Vineyard reservoir in water-poor Carneros.

CHAPTER NOTES

[1] Fred Kunkel and J.E. Upson, "Geological and Ground Water in Napa and Sonoma Valleys, Napa and Sonoma Counties, California," in Geological Survey Water Supply Paper 1495. USGS, 1960: p. 37.

[2] The names of watersheds in this section have been withheld to protect the privacy of individuals living or working in the areas where water supply or fish health has been questioned or compromised.

[3] Friends of the Napa River expected public release of this study by December 2001.

[4] CALFED is a coalition of state and federal agencies that, together with non-government stakeholders, is restoring the San Francisco Bay and Delta ecosystems.

[5] State Water Resources Control Board. 1990 Water Quality Assessment [for the State of California]. Sacramento: SWRCB, Division of Water Quality, Resolution No. 90-33, 1990.

[6] "Appendix D: Vineyard Heaters Management Plan," in *Napa River Watershed Owner's Manual: An Integrated Resource Management Plan.*

PHOTO CREDITS

Juliane Poirier Locke: pages 42–43, 44 *top*, 46 *bottom*, 48, 49, 50, 51 *top*, 52, 55, 56 *top*, 59
Vicki Kemmerer: pages 56 *bottom*, 57, 58
Dept. of Fish and Game: page 45
Tom Wilson: page 51 *bottom*
Collection of Tom Wilson: page 44 *bottom*
Collection of Robert Adams: page 53
Astrid Bock-Foster: pages 46 *top*, 47

Chapter Two: Water

Chapter Three

▲ A no-spray swale runs through this Carneros vineyard, where grower Al Buckland keeps herbicide away from any surface water flowing toward San Pablo Bay.

Vineyards in the Watershed

Weeds

Eliminating high-risk herbicides

In California vineyards, more than two-dozen weed species compete with vines for water and nutrients. To kill the competition, growers apply some form of chemical herbicide to 90 percent of all vineyard acres in the state.[1]

But some growers are questioning the need for these products because chemical herbicides can damage soil structure and soil microbes, pose a potential threat to worker safety, persist in the environment and may harm wildlife, water quality and human health.

In Napa County, most vineyard land gets some treatment with herbicides. Growers striving to farm sustainably fall anywhere on a graduated scale of improved practices that include switching from high-risk to low-risk herbicides, adapting application tactics to minimize herbicide product and, ultimately, eliminating chemical herbicides altogether.

Chapter Three *Weeds*

Along with ethical reasons, there are practical incentives for growers to re-think current herbicide use.

Groundwater at risk

Some of the push to change weed control practices comes from growing public awareness of what harm pesticides can do to the health of workers, neighbors, water quality and other elements of the watershed environment.

Along with ethical reasons, there are practical incentives for growers to re-think current herbicide use. Under the 1996 Food Quality Protection Act (FQPA), five of the eight herbicides used by Napa County growers are chemicals listed as the highest priority for environmental and heath risks *(see chart, opposite).*

One of the most commonly used herbicides in the county, simazine, has become a public enemy in the past few years. California well monitoring studies completed in 1998 showed that groundwater wells in agricultural counties were contaminated where pesticides including simazine had leached through predominantly sandy soils.[2]

Simazine earns infamy

No vineyard pesticides were detected in the three wells monitored in Napa County, possibly because Napa's predominantly clay soil makes travel difficult for pre-emergence herbicides. Yet under certain soil conditions these herbicides might travel into local groundwater. In some watersheds, because movement is so slow, it can take hundreds of years for contaminants to leave groundwater via natural processes.[3]

Two herbicides used by Napa winegrowers for weed control, simazine and diuron, have been found in groundwater in Contra Costa, Fresno and Tulare counties. During 1999, 224 pounds of diuron were used in Napa County vineyards; in the same year, growers applied 10,600 pounds of simazine, making it the second most widely used herbicide in Napa County, by weight.[4]

Known by brand names including Princep®, simazine is a pre-emergence herbicide—applied to the soil before the weeds germinate and remaining active as a weed seed killer in the soil. The soil life of a pre-emergence herbicide varies according to a number of factors.[5] Simazine is toxic to fish and wildlife[6] and has been listed as a "bad actor" herbicide by the public watchdog organization Pesticide Action Network North America (PANNA).[7]

Currently only 10 percent of local winegrowers use simazine, applying it to about 12 percent of the total vineyard acres in Napa County. Those percentages may be reduced through grower-driven educational programs promoting alternatives to conventional farming practices.

Continued on page 67

Herbicides used in Napa County, 2000

Herbicide	Pounds applied	Acres treated*	Moved into groundwater under legal farming use	Potential for movement to groundwater, according to DPR**	FQPA research priority***	Public watchdog organization rating****
Glyphosate (RoundUp®)	27,448	29,060		no	Priority II	
Simazine (Princep®)	10,579	4,219	Found in California groundwater	yes	Priority I	"Bad actor"
Oryzalin (Surflan®)	8,934	4,069		yes	Priority I	
Oxyfluorfen (Goal®)	8,130	10,792		no	Priority I	
Paraquat dichloride	745	782		no	Priority I	"Bad actor"
Norflurazon (Solicam®)	331	154	Found in California groundwater	yes	Priority II	"Bad actor"
Diuron	224	129	Found in California groundwater	yes	Priority II	"Bad actor"
2,4-D	58	43	Found in U.S. surface waters	No (but has been found in wells in other states)	Priority I	"Bad actor"

* The number of acres treated includes multiple treatments. For example, if a 10-acre vineyard is treated twice in a year, it counts as 20 acres in this total. Consequently, this category does not represent actual land area treated.

** California Department of Pesticide Regulation, Ground Water Protection Program

*** Under the Food Quality Protection Act of 1996, pesticides will be phased out according to priority status assigned to each chemical product.

**** Pesticide Action Network North America, a public interest watchdog group, assigns the "bad actor" rating to California registered pesticides based on whether they are "acute poisons, carcinogens, reproductive or developmental toxicants, neurotoxins or groundwater contaminants." *Hooked on Poison: Pesticide use in California 1991–1998*, by Pesticide Action Network North America. San Francisco: Californians for Pesticide Reform, 2000.

Case Study: Mitchell's flocks—the living lawnmowers of Carneros
Napa River/Huichica Creek watershed

▲ Mitchell Klug

Some growers contend that weeds stake more claim on a vineyard in Carneros than in any other part of Napa County. In this Carneros case study, livestock participate in an evolving management plan for sustainable weed control.

Mitchell Klug has no training as a shepherd, but he knows where his sheep will be this winter in Carneros—eating weeds at the vineyards he manages for Robert Mondavi Winery.

"Weeds are our biggest problem here in Carneros," Klug said. As one alternative to herbicides in established vineyards, Klug leased sheep in November 2000 for a weed-eating mission that lasted through March 2001. The flock of 35 sheep reduced weed pressure on 40 acres. This winter 1,000 sheep will be controlling weeds on 134 acres.

"We want to get the animals to do the work of machines," Klug said. The barnyard-to-vineyard recruits he dubs "living lawnmowers" include more than sheep. Geese—purchased when they were goslings and fed on target weed species—have turned out to be voracious weeders. The whole gaggle get fenced into quarter-acre sections at a time, gradually eating their way through 10 acres a year. The geese also dine on plants around the reservoir, including PD plants—host plants for Pierce's disease. A pair of ducks recruited to aid the geese tag along as part of the work force. "We have a Dr. Doolittle farm," Klug quipped. "All we need is Eddie Murphy."

While the Carneros menagerie is experimental, there have been goats keeping down the weeds at Mondavi's Stags Leap location for the past eight years. On that property, 35 goats graze around the Mondavi home and the reservoir while 12 cows graze the fence lines, providing fire protection and PD plant control on 200 acres. The grazers contribute to soil nutrient cycling and microbial activity as an added benefit.

▲ These "living lawnmowers" are one of Klug's several non-chemical weed control experiments.

◀ When enough weeds have been eaten for the day, off-duty sheep and geese flock to this pond at Mondavi Carneros ranch, where setbacks total six acres. The preserved area provides a natural flyway to nearby marshlands for passing waterfowl. The pond allows weed-eating fowl on the ranch to escape into water when threatened by predators such as coyotes.

Animal assistants on the Carneros ranch work in tandem with a planting strategy. Vine rows—the area directly under the vines—are planted to cover crops as another non-chemical weed control strategy. Zorro fescue is used for a number of reasons.

"It's a good grass for removing other weed species," said Klug. "It's low stature, is a prolific seeder and goes dormant really early. The seeds are small, so it's not desirable food for rodents." Klug said the fescue grows short enough that it will not interfere with the vines even if not mowed; however, it does get mowed for fire safety and for better working conditions, since the surface of the dried grass on hillside vineyards can be too slick for workers to stand on safely.

Weed control in the established blocks is also done by manual and mechanical means including mowing, hoe plowing, knife weeding and hand weeding.

Knife weeding has been effective controlling annuals that appear after October and November rains. After the knife weeding, the sheep keep weeds under control until bud break in the spring.

Klug is pleased that at the combined Mondavi vineyards in Napa County, no herbicide was used on 1,400 acres the past two years. But to control weeds on the developing vineyards, the contact herbicide RoundUp® is still used.

RoundUp®, a brand name for glyphosate, is thought to be among the most environmentally friendly chemical herbicides because it can, under some circumstances, biodegrade in half a day. Yet some growers believe that a substitution can and should be made for even the least harmful herbicides. According to Klug, he and everyone else trying to grow sustainably need to take the next step and ask, "How do we get rid of RoundUp®?" ■

Chapter Three: Weeds

Chapter Three *Weeds*

▲ Jon Kanagy

Learning from a nemesis

Visiting your weeds—those plants growing where you'd rather they didn't—can be educational.

"Weeds can tell you something about your soil," explained Jon Kanagy, of Nord Coast Vineyard Service. "Sorrel, for example, tends to inhabit wet places, so its presence may indicate poor drainage." Other weeds may indicate nutrient richness or deficiencies.

It's important, Kanagy stressed, for growers to know whether their weeds are annual, perennial, winter or summer weeds, and whether they actually compete with the vines. Some weeds, such as bindweed, are active and growing on the same time schedule as grape vines, directly competing for water and nutrients. But not all weeds compete. In fact, weeds used as cover crops can actually be good for the vine *(see Chapter 5)*.

Weeds that move in quickly and become abundant, Kanagy explained, are what ecologists call R-selected species, associated with primary succession.

Before agriculture and before cattle grazing, the sites where there are now vineyards were likely free of these weeds because they were probably covered in native grasses. These California grasses are examples of K-selected species associated with secondary succession, or population equilibrium.

"After a natural disturbance like a landslide, the bare earth tends to be colonized first by R-selected species," he explained. "They don't need much to prosper. They are adapted to reproducing in large numbers very quickly."

Like the English of old, these species are ambitious colonizers. Red-root pigweed, for example, invests its resources in a very high number of seeds per individual plant—many thousands—with the idea that a few of them will survive and take hold.

When farmers are continually disturbing the ground, they are fighting succession, Kanagy said, "creating a space where only R-selected species can survive." Some farmers are experimenting with K-species, such as California native perennial grasses, in cover crops as a way to keep out those aggressive R-selected species.

Since it is likely that agriculture will keep disturbing the soil to some extent, and R-selected species will continue their relentless attempts to colonize, the relationship between farmers and weeds seems destined to continue. Whether the relationship presents a learning opportunity is entirely up to the grower. ■

Continued from page 62

From high-risk to low-risk

Roughly 29,000 acres of Napa County vineyard land is treated with the contact herbicide glyphosate, sold under brand names including RoundUp®—the same chemical commonly used in residential yards and gardens. Glyphosate is applied directly to plants and kills weeds systemically. Glyphosate does not present as high a toxicity risk to the environment nor does it persist as long in the environment as simazine.[8]

Shifting away from pre-emergence herbicides can, for some, mean moving toward better weed control strategies and better farming. Kelly Maher, director of vineyard operations for Domain Chandon, said sustainable weed control is more challenging and requires more kinds of knowledge than the chemical strategy he described as once being the norm. In the fall, he said, conventional farmers would get out on their tractors and spray their pre-emergence "cocktail." End of strategy. The so-called cocktail was typically a mixture of pre-emergence, contact and systemic herbicides.

Kick habit, farm better

"When you get off the chemical dependency," Maher said, "as soon as you switch from pre-emergents, you have to

What weeds want

Some weeds, if you study them closely enough, can act as environmental indicators. The uninvited plant is not there to torment vineyard managers, but to fill an environmental vacuum. By thriving on what's in a particular soil, weeds can offer clues about soil health and balance.

- Mustard plants like to mine for nutrients in well-tilled soil that is usually low in nitrogen. So if you're trying to fix nitrogen with a cover crop, beware that mustard doesn't compete with it.
- The morning glory—or bindweed, named for its throttling tendencies—likes to grow in anaerobic soils, often low in calcium, phosphorus, potassium and pH. A sun-loving colonizer, it can be discouraged by the shade from an over story crop just two to three inches tall.
- Where curly dock and sour dock are growing, the soil may be anaerobic, acid, poorly drained and high in magnesium. Docks might be growing over a compacted section of the vineyard.
- Thistles thrive in disturbed, high-clay soils that are low in organic matter. They are a tap rooted plant, mining nutrients from deeper soil.

(Bulleted text adapted from a presentation by Kirk Grace, August 10, 1999, Napa Valley College.) ∎

Chapter Three *Weeds*

Some non-chemical weed controls

> *"When you lose your chemical dependency... you're out in your vineyard more and you're going to be more aware."*
>
> —Kelly Maher, Domaine Chandon

learn your weeds." He observed that growers who have been using pre-emergence herbicides for 20 years have never seen their weeds, so it's "no wonder they have a problem identifying them."

Maher said the vineyards he runs for Domaine Chandon "got off the cocktail" in 1998. He has explored alternatives to herbicides to improve worker safety and soil health, and also to be more in tune with a systems approach to farming. In the systems approach, the "cheap" herbicide cocktail, at $25 an acre, is neither environmentally nor socially affordable.

"Mow and throw is probably the future for us," said Maher, referring to the practice of using cover crop mowings as mulch in the vine rows. Timing and weed identification, he stressed, are the key factors in a sustainable weed control program.

There's a hidden advantage to all the extra work it takes to control weeds sustainably, according to Maher. "When you lose your chemical dependency and need to identify your weeds," he said, "you're out in your vineyard more and you're going to be more aware. You start noticing other things." Maher believes that spending more time in the vineyard has vastly improved his skills as a farmer.

"A farmer who's more in tune with his vineyard," said Maher, "is a better farmer."

▲ At Joseph Phelps vineyards in St. Helena, Philippe Pessereau measures the relative height of weeds after three in-row flaming treatments. Timing is critical for the effective use of a weed flamer.

▲ At a Hess Collection vineyard on Mt. Veeder, this hydraulically driven and controlled mower (operated manually here) goes in and around the vines to cut weeds. The mower adjusts for upper and lower terraces—the toughest conditions for mechanized weed control.

▲ Preserving the California golden poppy for aesthetic value, Alfonso Valencia and Herman Hernandez shovel-weed Joseph Phelps Vineyards' Backus Ranch near Oakville. The $120 price tag on a bottle of wine from this vineyard helps make the labor costs sustainable.

▲ Poised atop a historic stone structure, Ferdinand the Bull—a non-conformist character out of children's literature—offers whimsy and wind direction.

▲ ▶ At Frog's Leap's Galleron Ranch in Rutherford, Gilberto "Gabby" Cisneros operates a hoe plow to cut in-row weeds away from young vines. The job requires a highly skilled operator in order to protect vines from possible tractor damage.

Chapter Three: Weeds

Horse-powered weed control, vineyard management

▶ Montana, a Belgian draft horse, pulls grapes to the staging area during harvest.

▶ Bob and Duke pose with Eric Grigsby. These Belgian draft horses are trained to work in the vineyard.

A segment of the 22-acre Rocca Family Vineyards in Yountville has been exclusively horse-worked for the past two years, an experiment with reducing weeds, tractor noise, diesel exhaust, soil compaction, herbicide use and—perhaps most importantly for owner Eric Grigsby—increasing the fun of farm work.

Working horses in the vineyard is "way more fun than driving a tractor," said Grigsby, whose vineyard foreman Sergio Melgoza works three Belgian draft horses to cultivate, control weeds and even pull grape loads at harvest.

Grigsby is testing a horse-drawn weed control device, "something like a hoe plow," he explained. It has a retractable arm instead of a cultivator and spinning tines that stir the dirt. He has eliminated the use of simazine and Goal® for weed control and currently uses only RoundUp®. He would like to control weeds without any herbicides and hopes that the new weeding tool proves effective.

Grigsby calculates that a 2,000-pound workhorse is far less likely to cause soil compaction than a 3-ton tractor. For workers who prefer not to hear the sound of loud farm machinery or inhale the byproducts of a combustion engine, working horses in the vineyard may improve working conditions for employees.

▲ This ride-on, horse drawn rotary harrow replaces a tractor at the Rocca Family Vineyards.

According to Grigsby, tractor noises, diesel odors and exhaust are replaced with peaceful quiet and clean air when horses are used instead of machinery. "All you can hear are the guys talking," Grigsby said, "and the sound of the horses breathing."

When friend and grower Julie Johnson-Williams of Williams Ranch pointed out that horses might want to snack on grapes at harvest, Grigsby was amused rather than deterred. He is hoping that in time other growers will be trying horse-powered weed control and vineyard management so they can experiment, trade ideas and solve these problems together. ■

No more silver bullets

"One tenet of sustainable farming," said a Carneros winegrower with 20 years experience, "is not to formula farm, but to farm according to site."

Some growers, in an attempt to farm sustainably, are seeing weeds in a new light, discovering how weeds can indicate soil condition. Others are gaining new awareness of how chemical inputs affect the essential functions of soil microbial communities.

Farmers working a sustainable program are trying to get away from "silver bullet" solutions and experimenting with a variety of tactics to battle weeds in the vineyard: switching from high-risk to low-risk herbicides, spraying less weed killer, spraying no weed killer, shovel weeding, planting in-row cover crops, setting livestock into the vineyard to graze and testing mechanical weed-control.

The success of these strategies depends on site-specific factors, but one generalization can be safely made: those farming a sustainable program do not use chemical "cocktails" to solve weed problems.

CHAPTER NOTES

[1] Jenny Broom, PhD., "California Winegrape Pest Profile, Use, and Research Needs Under 1996 Food Quality Protection Act," in *Sustainable Agriculture,* Vol. 12, No. 1. (Winter/Spring 2000): pp. 12-13.

[2] California Department of Pesticide Regulations, *Sampling for Pesticide Residues in California Well Water: 1998 Update of the Well Inventory Database for Sampling Results Reported From July 1, 1997 through June 30, 1998:* pg. 21.

[3] "Pesticide Safety" in *Grape Pest Management,* 2nd ed., University of California Division of Agriculture and Natural Resources, 1992: pg. 359.

[4] Pesticide use records, Napa County Agricultural Commissioner's office.

[5] *Herbicide Handbook,* 7th ed., Weed Society of America, 1994: pp. 270-272.

[6] Rob Forster et al., *California Wildlife and Pesticides,* California Department of Fish and Game, Pesticide Investigations Unit, March 1997: pg. 35.

[7] Pesticide Action Network North America, toxicity database, 2001.

[8] *Herbicide Handbook*: pp. 149-152.

PHOTO CREDITS
Terence Ford: page 64
Juliane Poirier Locke: pages 60–61, 65, 66, 68 *top*, 69
NRCS: page 67
Richard Camera: page 68 *bottom*
Kathy Kellebrew: page 70 *bottom right*, 71
Nate Grigsby-Rocca: page 70 *left*
Eric Grigsby: page 70 *top right*

Chapter Four

▲ Healthy creeks provide food and shelter for great blue herons and other wildlife species.

Wildlife

Protecting habitat, courting diversity on the farm

The fact that Napa County remains the only rural county in the nine-county Bay Area, and not the bedroom community it was targeted to become, is due largely to the efforts of the agricultural community in creating and defending the Agricultural Preserve.[1]

This zoning strategy has prevented wholesale urbanization, assisted in preserving the unique character of the local community, and perhaps helped to slow—but not stop—the loss of Napa County's wildlife habitat.

Habitat is the combination of area, cover, food and water needed to support life for wildlife species. Within the physical setting of the habitat, plants and animals have complex, interdependent relationships that have evolved over time and upon which they depend for survival.

Chapter Four *Wildlife*

Of the 52 habitat types distributed throughout California, Napa County is host to 26, including natural woodlands, wetlands and grasslands. Also present here, and expanding, are urban and vineyard habitats.[2] As people continue to develop land—whether for houses, industrial buildings, ranches or vineyards—wildlife habitat is diminished.

▼ Spotted owl

Fragmenting home turf

This trend is not unique to Napa. Wildlife habitat—and the biodiversity that characterizes it—is in crisis everywhere, due largely to fragmentation.

Developments of any kind interrupt natural systems by segmenting land areas. When the segments are too small, many species of animals and plants quickly disappear, frequently to extinction.

Habitat for the spotted owl in Angwin, for example, has been fragmented by urbanization and vineyard development. Ted Wooster, a retired Department of Fish and Game (DFG) biologist who has been studying spotted owl populations in Napa County since 1976, said that small patches of remaining habitat had recently supported three pairs of owls in Angwin. "The third pair is gone now," said Wooster.

Federally listed as a threatened species, the spotted owl still resides in woodlands on the east side of the Valley near Calistoga and on Mt. Veeder to the west.

Case study: Managing for wildlife and water quality
Napa River/Huichica Creek/Carneros Creek watershed

At Hudson Vineyards in Carneros, minimal development, generous setbacks and reforestation efforts combine to foster clean water, protect wildlife and sustain a working balance of ecology, beauty and business interests.

"Agriculture is industrial. It ain't bird watching," said Lee Hudson, of Hudson Vineyards in Carneros. "But you can try to do everything you can to promote birds and wildlife."

For Hudson, promoting wildlife includes establishing riparian setbacks that exceed regulatory demands. On a cool Carneros morning in spring, Hudson drove his Jeep around the 180-acre vineyard site where he's been farming since 1981—a hilly landscape bordered by oak woodlands, oak grasslands and a reservoir dotted with water birds. The vineyards don't crowd the waterways on this land. From vine to creek, the vineyards are set back an average of 75 feet.

Both Huichica and Carneros creeks flow through the Hudson property, where only 10 percent of the property has been developed as vineyards. "I measure my setback from the end of the turn-around," said Hudson (a turn-around is the area in which a tractor can exit a row and enter the next row). "I add 25 feet to a standard setback, and if there's any more reason that drives me, I'll even add more than that."

One reason Hudson will increase the setback area is the presence of a steep bank or big trees. As a rule, Hudson keeps farming operations a minimum of 30 to 40 feet from the drip line of a big tree.

These setbacks are a serious economic decision. "Whatever sustainability is," he said, "it's being responsible and at the same time staying in business." Although he concedes that there are "no business risk and no production costs" on the land set aside for wildlife, there is also no possibility for income.

"One acre is a tremendous loss of revenue," Hudson explained. "Say it's four tons—that's 600 gallons of wine at 2.24 gallons per case. That's 250 cases of wine, with a margin of $50 a case." Hudson figures that an acre given to setbacks can therefore represent up to $14,000 in lost revenue. "That's a serious donation," Hudson said. But the value of a healthy riparian corridor, for sustainability, may have even greater value in terms of the environmental services it provides.

Riparian corridors, such as those created by setbacks, support both wildlife diversity and water quality. As some farmers have discovered, a

▲ Lee Hudson has a keen interest in sustaining both business and wildlife.

"Whatever sustainability is, it's being responsible and at the same time staying in business."

—Lee Hudson

healthy riparian area acts as a buffer between the vineyard and the waterways, filtering out sediments that might otherwise pollute the watershed. "It mitigates the impacts of agriculture," Hudson explained, "and allows the water to leave the property as clean as when it fell from the sky."

Water quality is a big concern at Hudson Vineyards, where the agricultural land is adjacent to the habitat of the California freshwater shrimp (*Syncaris pacifica*), an endangered species. Hudson was among the growers who collaborated in the first watershed stewardship group in Napa County (see page 149).

"The Huichica Creek (stewardship group) process was a positive philosophical adjustment for me," Hudson said, "to look at my property as part of the watershed." He now belongs to the Carneros Creek stewardship group, but does not use his membership in stewardship groups to gloss over the weak points in his farming plan.

"My most difficult (farming) task is reducing my dependence on herbicides," said Hudson. "Everything else is relatively simple." Hudson, who still uses pre-emergence herbicides for weed control and is investigating alternatives, says his goal is to reduce herbicide use by 50 to 75 percent over the next three years.

Meanwhile, because the farming area is only a small portion of the total property, his land management plan includes reforestation. To help enhance the oak woodland surrounding the vineyards and add to the biodiversity of the place, Hudson and his crews have planted buckeye, blue oak, live oak, valley oak and manzanita. They collect acorns from oaks on the property and use them to start seedlings, which are planted and protected with plastic sleeves.

From a business standpoint, Hudson says, there is no direct benefit from planting trees. He does it for "aesthetic enjoyment and calming the soul." For numerous wildlife species, the trees offer more practical benefits, including the means for survival. ∎

▼ The vineyard setbacks on Hudson's property are among the most generous in the county.

Although the spotted owl is almost a celebrity species, hundreds of thousands of other species—both plants and animals—have been pushed to extinction without getting their names in the paper.[3]

Every species needs protection

People tend to favor protecting those species they find attractive or those that inspire sympathy, a phenomenon described as the Bambi syndrome.[4] But all species in an ecosystem are important for the job they do or the food they provide, even down to the tiniest soil micro-organisms essential for plant survival. We humans owe our existence in part to the life cycles of rather humble and unattractive species.[5]

Some species are mentioned more frequently than others in discussions about the environment because they are what scientists call indicator species. An indicator species is to biodiversity what the canary in the coal mine was to miners.[6] When the survival needs of an indicator species are met, there is a good chance that the survival needs of numerous other species will be met also.

▲ Restored riparian corridor on the Napa River.

Property and stewardship

For many reasons—including concern for preserving biodiversity—landowners practice stewardship voluntarily. Lee Hudson, owner of Hudson Vineyards *(see page 75),* for example, voluntarily establishes generous riparian setbacks and plants trees to augment the woodland habitat on his property. Other examples of voluntary land stewardship are found throughout this book.

When landowners do not practice habitat stewardship on their own, public concern about the environment pressures them to do it anyway. Fines for poor stewardship are increasing. For one Sacramento area vineyard developer, altering two acres of habitat critical to the survival of an endan-

Chapter Four: Wildlife

Chapter Four *Wildlife*

◀ Blacktail deer and ringtail (below) are among the numerous wildlife species that use riparian corridors.

Riparian corridors—the wider, the better

People interested in sustaining biodiversity in the long term are "going to need to think about wider corridors," said wildlife researcher Jodi Hilty, whose UC Berkeley study looked at animals active along waterway habitats.

Conducted over a five-year period—with fieldwork done during three consecutive summers in Sonoma County—Hilty's study employed fixed video cameras that taped wildlife activities in riparian corridors.

Her research indicated that the area and quality of habitat on either side of a waterway corresponded with the level of biodiversity and kinds of animal activity found at each site. "One important condition for biodiversity," said Hilty, "is space."

The adequate width of a riparian corridor is hotly debated in Napa County and elsewhere. For the purposes of the study, wide corridors were defined as anything over 29 meters on each side of the waterway; narrow corridors were between 10 and 29 meters on each side of the waterway; and denuded creeks were between 0 and 10 meters on each side of the waterway.

In the denuded creeks, the dominant species were possums, raccoons and domestic cats—all highly active. Sometimes up to six domestic cats were sighted in a narrow corridor, whereas in wide corridors there was never more than one cat observed.

The presence of cats and raccoons "becomes very relevant for birds," Hilty says, because, like cats, raccoons are nest and bird predators, and they do well in human disturbed areas. High numbers of raccoons and domestic cats, according to Hilty, result in "a really hammered bird population." Birds are often pests in vineyards, but they can be beneficial as well.

Winegrowers interested in the insect-eating qualities of birds can encourage bird populations by maintaining wide riparian corridors for habitat, according to Hilty. Birds such as starlings, which are often unwelcome in the vineyard, will come whether or not there is a riparian corridor. A healthy riparian canopy "won't increase your populations of the bad species, but it will enhance populations of the good birds," Hilty said. "Even birds that are pests to grapes later feed their young on insects."

Bird species in wide riparian corridors include warblers, flycatchers and woodpeckers. These wide corridors also draw more native species such as coyote, bobcat, gray fox and skunk. To stabilize a diverse wildlife corridor, Hilty's research suggests that riparian corridors be replete with vegetation and over 29 meters on each side of a waterway. ■

gered species resulted in fines of $1.5 million.[7]

Vineyard developments planned with respect to watershed processes and the protection of public waterways can result in many benefits, including positive community relations, business profits and resource sustainability.

John Williams, owner of Frog's Leap Winery and Vineyards, restored a section of the Napa River adjacent Galleron Ranch in Rutherford.

The site, once denuded and plagued by bank failure, is now stabilized by trees and shrubs and frequently toured by visitors interested in watershed

Wildlife fencing—knowing where and when to use it

"Don't put in fencing if you don't need it," says Allan Buckmann, Department of Fish and Game (DFG) wildlife biologist. "The perimeter fencing people are putting up around entire properties is killing off lots of animals indirectly."

Fencing just around the vineyard block is sufficient, according to Buckmann, who advises property owners not to develop an entire site, but to plan developments around natural constraints including the travel routes of wildlife.

Consultants can be hired to survey a property's wildlife resources and to map the natural constraints so owners can know what's on their site. "Most people are afraid of what they're going to be told, but a lot of times those sensitive features tend to be really small and often are in convenient places," said Buckmann. "If you have sensitive species, you can plan around them."

Once sensitive resources have been identified, Buckmann explained, DFG personnel will "try to help them make a plan that's good for agriculture and for wildlife, so we get the best of both worlds."

Tips for correct wildlife fencing

Open corridors	**Temporary fencing**	**Permanent fencing**	**Fence, don't shoot**
All the waterway corridors should be isolated as corridors, and no fencing should be placed across or within those corridors.	Since young vines are most attractive to wildlife, a temporary electric fence can be used in most cases. However, permanent fencing is necessary to protect vines from wildlife at some sites.	Fencing should be at least 8 feet high so deer can't jump over it and limited to vineyard blocks. In blocks more than one-half mile wide, a corridor should be installed. Gates should go in the corner rather than in the middle of a fence.	The DFG expects landowners to use fencing where wildlife can enter the vineyard. Landowners cannot expect to rely on DFG depredation permits to shoot the wildlife that becomes a problem in unfenced vineyards.

Chapter Four *Wildlife*

Animals can take full advantage of their habitat range, except when blocked by fencing.

stewardship. The restored corridor now offers water-cooling shade required by many fish species, insurance against erosion of vineyard land, wildlife habitat and a picnic area for workers and family members. Williams found that the riparian restoration project was good for wildlife and good for business.

Diversity equals stability

Many farmers recognize that wildlife deserves a place on the farm as much for the sake of the farm as for the sake of the wildlife. Healthy wildlife habitats improve the aesthetic value of the land and contribute to the long-term sustainability of the vineyard. Wildlife habitat provides services such as water filtering, pest reduction, soil health, erosion control and beauty.

In managing vineyards for wildlife, encouraging biodiversity is key. This means encouraging the number and kinds of organisms that can make a home there. In nature, diversity creates stability, while unstable ecosystems often result in population imbalances, including pest infestation *(see Chapter 6)*.

Although biodiversity in and around a farm has never been formally assessed in California vineyards, one organization has developed a checklist for biodiversity linkages[8] to measure a property's biodiversity in categories of composition, structure and function. A similar checklist might be adapted as a tool for land managers to evaluate Napa County properties and farms.

Cover crop buffers

Certain cover crops, grown in the form of hedgerows or insectaries, also benefit the vineyard through insect control and increased biodiversity. These crops, in addition to in-row and between-row cover crops, can be grown as buffers between the vineyard and the natural habitat surrounding the vineyard.

▼ Proper fencing can allow wild turkeys access to waterways and can keep them out of the vineyard.

Conservation easements—landowners and wildlife can benefit

"The wolf of development is ever at the door of this county," said Jon Hoffnagle, director of the Land Trust of Napa County, a non-profit corporation. While urban development is a pressure on lands everywhere in California, Napa County's unique approach to conservation preserves farmland and open space while benefiting landowners, farmers and wildlife.

To keep the wolf at bay, preserve the area's rural quality, reduce income taxes and help keep family farms in the family—not sold to pay inheritance taxes—Hoffnagle thinks landowners ought to consider conservation easements.

Local landowners working with the Land Trust have preserved 25,000 acres—5 percent of the lands in Napa County: 12,000 acres were made conservation easements; 3,000 acres were sold to the Land Trust; and 10,000 acres were purchased and held by the Land Trust until the state could buy the acreage for parkland.

A conservation easement is a restriction of what a piece of land can be used for—it is not a loss of ownership. Land under easement can be gifted to heirs or sold. An easement of this kind does not give public access to a property.

About a third of the land donors participating in this conservation program are farmers who want the land kept in agricultural uses. The rest are owners of open space land they want to keep undeveloped. Every easement agreement is unique, a collaboration between the landowner and the Land Trust.

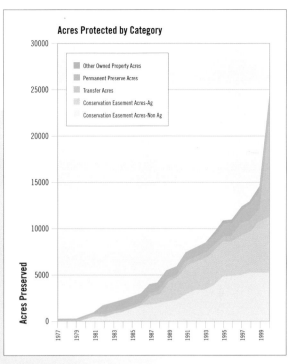

"Conservation easements underscore what the owner wants to do with the property," Hoffnagle explained, citing Joseph and Lois Phelps' conservation easement of 480 acres, including 130 vineyard acres, resulting in a multi-million dollar tax deduction and a guarantee that the land will never be developed for housing. Another landowner used easements to guarantee that his property would go to his children and not have to be sold to pay inheritance taxes.

With what Hoffnagle calls the "graying of land ownership" in Napa County, there may be a need to create conservation easements sooner than later. The Agricultural Preserve, a zoning ordinance that protects farmland from non-agricultural uses, is set to expire in 2020. Hoffnagle believes that if the political environment shifts toward growth and development, land in conservation easements is the best insurance against those urban development wolves. ■

Chapter Four *Wildlife*

Cover crops of native grasses used as buffers create a transition zone, increasing diversity, providing cover and habitat and acting as filters to improve water quality.

Buffers trap both nutrients and pesticides that get washed off agricultural lands in the rain, preventing them from entering the waterways. Studies of other agricultural areas show that up to 10 percent of an applied pesticide can be washed off a crop by heavy rain, and the water-filtering property of buffers is "the most important factor in trapping these pesticides."[9] Buffers also trap sediment, slowing erosion.

▼ A "grandfather" oak preserved on the Valley floor near Yountville.

Leave those trees

Trees, too, provide both watershed and vineyard services. Keeping as many trees as possible in any vineyard development plan is critical. Adina Merenlender, a UC Berkeley Cooperative Extension woodlands specialist, says, "Keeping vineyard development outside the drip line of mature trees" will help both the trees and the grape vines survive in a healthy state. Leaving trees uncut can, in some circumstances, help prevent the spread of oak root fungus (Armillaria root rot).[10]

Preserving woodlands also means preserving wildlife. An oak woodland, for example, provides "critical habitat for approximately 2,000 plants, 160 bird, 80 mammal, 80 amphibian and reptile and 5,000 insect species."[11]

Trees also offer beauty, shade and cover for humans and animals, filtration of sediments and control of erosion. Some landowners in Napa and other counties with dwindling oak woodland habitat are voluntarily collecting acorns, germinating them, planting the seedlings and protecting young trees with plastic sleeves until the trees are well established.

Riparian corridors

Trees and shrubs growing along waterways are especially important to wildlife. As part of the riparian habitat, these trees and shrubs—in addition to controlling erosion—provide shade to cool the water and suit the temperature needs of fish, offer food and living space for on-site creatures and create cover for those on the move.

Animals traveling down from the hills, unless they can fly, need continuous cover to avoid predators. In healthy riparian corridors, the cover, shade and moisture support the migration of deer, mountain lions, ringtails, weasels and river otters. Weasels, in particular, are beneficial to grape growers because they prey on rodents.

Landowners who replant these denuded creek sites with native vegetation help support a continuous habitat and contribute to wildlife biodiversity. By restoring healthy riparian canopy, these landowners are providing habitat for the bluebirds, swallows and bats that eat vineyard insects.

Restoring riparian corridors means animals can take full advantage of their habitat range, except when blocked by fencing *(see sidebar, page 79)*.

Farming around wetlands and vernal pools

It's not only law but common sense to avoid planting vineyards in or near wetlands, since such a practice is bad for grapevines as well as wetland wildlife habitat.

Vernal pools are seasonal wetlands with cycles that are not always predictable. It often takes a trained eye to detect the presence of a vernal pool when the area is going through a dry stage, but like regular wetlands, vernal pools contribute to water quality, support rich biodiversity and are managed strictly as non-farmed areas.

Wildlife and winegrapes

New vineyard development disturbs and depletes wildlife habitat. But sustainable development practices preserve as much habitat as possible —leaving as many trees as possible and providing generous setbacks from the waterways, wetlands and vernal pools.[12]

Because sustainable land management practices focus equally on both the farmed and non-farmed areas of a vineyard, grassland and "weeds" are left intact wherever possible, denuded creeks are replanted with native vegetation and strategically placed wildlife fencing includes wildlife corridors.

Animals traveling down from the hills, unless they can fly, need continuous cover to avoid predators.

Chapter Four: Wildlife

Chapter Four *Wildlife*

▲ Huichica Creek on the RCD Demonstration Vineyard site, before and after. The creek was restored with native vegetation and impacts of livestock and human use were eliminated.

▶ A wetland that had previously been graded and drained was restored and is now managed as non-farmed habitat within a 2-acre portion of the property.

The RCD Demonstration Vineyard

The RCD Sustainable Agriculture Demonstration Vineyard, located in the Carneros region, is managed for wildlife habitat, watershed stewardship and commercial profit. Developed as a working, long-term project, the staff refer to it as a "come on down and kick the tires" operation. No replicated research plots are to be found within the 18-acre parcel, where visitors can observe a vineyard managed in an environmentally sensitive area that produces high-quality grapes and enhanced wildlife habitat along the northern fringe of the Napa-Sonoma marshlands.

The vineyard property was a dairy pasture 20 years ago, where cows used to cross through the creek twice a day on their way to and from the milking barns. Beginning in 1991, RCD and NRCS staff and volunteers planted native vegetation, stabilized stream banks, and set aside non-farmed riparian management zones as part of a land restoration program.

A wetland on the property that had been graded and drained has now been restored and is managed as spring waterfowl nesting habitat.

Native cover crops are maintained in the vineyard, which is not cultivated in order to maintain soil health and minimize loss of carbon and organic matter. Aside from an occasional spot treatment of RoundUp®, the vineyard is not treated with herbicides. Weeds are mown and resident native grasses are encouraged to spread into non-farmed set-aside areas. ■

CHAPTER NOTES

[1] Volker Eisele. "Twenty-five Years of Farmland Protection in Napa County" in *California Farmland and Urban Pressures: Statewide and Regional Perspectives.* Medvitz, Albert G., Alvin D. Sokolow and Cathy Lemp (eds.). Davis: Agricultural Issues Center, Division of Agriculture and Natural Resources, University of California, 1999: pp. 103–123.

[2] Kenneth E. Mayer and William Laudenslayer, Jr. (eds.). *A Guide to Wildlife Habitats of California.* California Resources Agency, 1988.

[3] David Quammen. *The Song of the Dodo: Island Biogeography in an Age of Extinctions.* New York: Scribner, 1996.

[4] DFG plant biologist Ken Berg once explained that the public is less inclined to protect plants and animals that do not have "big Bambi eyes."

[5] For a humorous, factual treatment of this topic, see Gary Larson's book, *There's a Hair in My Dirt: A Worm's Story.* HarperCollins, 1998.

[6] Because canaries asphyxiate before humans do, these birds were routinely carried into mining shafts where the oxygen supply was often poor; when the canary showed signs of failure or died, it was a warning for workers to get out fast to save their own lives.

[7] Jill Goetz. "Wine Growers Face New Complexities." *California Agriculture,* Vol 54 No. 3 (May-June 2000): p. 6.

[8] This checklist is part of an unpublished, private study: May 2001.

[9] "Conservation Buffers to Reduce Pesticide Loss." USDA Natural Resources Conservation Service, March 2000: p.1.

[10] Adina M. Merenlender and Julia Crawford. "Vineyards in an Oak Landscape." Berkeley: University of California Division of Agriculture and Natural Resources Publication 21577: pp. 7-8.

[11] *Ibid.* p. 4

[12] Scientific support for this claim can be found in "Use of Riparian Corridors by Wildlife in the Oak Woodland Vineyard Landscape," by Jodi A. Hilty (2001): pp.19-21. Ph.D. dissertation, Berkeley: University of California.

PHOTO CREDITS
Hal Beral: page 85
Nick Elias: page 78 *top*
Juliane Poirier Locke: pages 75, 76, 77, 82
Mark Salvestrin: page 74
Robert Waldron: page 80
David Welling: page 78 *bottom*
A.J. Wool: page 72–73
Land Trust of Napa County (data) and Max Seabaugh (illustration): page 81
Unknown: page 84 *top*
Astrid Bock-Foster: page 84 *bottom center*

▲ Snowy egrets are among the water birds drawn to healthy riparian habitat and to wetlands left undisturbed.

Chapter Four: Wildlife

Chapter Five

▲ **Molate red fescue** is a California native grass used in cover cropping.

Cover Crops
Preserving and improving the soil

During the past few centuries, winegrowers took pride in clean cultivation around their crops, even on steep hillsides. Evidence of "properly" farmed land was the sight of bare, tilled earth surrounding the vines. Until recently, the bare-earth aesthetic was so unquestionably an indication of farming competence that when one grape grower in the Fresno area began using cover crops, his neighbors—observing what appeared to be weeds—assumed the farmer had died.[1]

Cover crops, planted in the middle of the vineyard and sometimes under the vines themselves, include grasses, clovers, peas and other small plants. Despite their relatively small stature, these plants are assets to the vineyard and to the environment. Perhaps most importantly, cover crops

Chapter Five *Cover Crops*

▲ Zorro annual fescue is favored for its superior growth characteristics, short stature and re-establishment vigor.

are the best erosion control tool available. On much of the agricultural land in Napa County, the bare-earth aesthetic has been abandoned in favor of cover cropping.

But the change didn't happen overnight; old practices die hard, and the humble cover crop had to prove itself.

When cover crops first appeared in Napa County vineyards, they weren't an immediate hit. All too frequently, farmers would give a token try and then disk them up after a season. Dennis Moore, a soil conservationist with the Soil Conservation Service (now the NRCS) during the 1970s, tried to get hillside growers interested in making a serious commitment to no-till cover crops for erosion control. It was a hard sell back then, since many growers didn't consider erosion a problem.

But a few growers were ahead of their time. Chappellet, Sterling and Christian Brothers were among the first to make cover crops a permanent feature in some of their hillside vineyards in 1978—more than 20 years before the conservation regulations made hillside cover crops mandatory.

An economic push

While there is no erosion control requirement for cover crops on the flatlands, grasses and peas made it to the Valley floor anyway, thanks—in part—to the price of gas.

During the 1970s, running tractors to maintain clean cultivation was getting expensive. DeWitt Garlock, working for Christian Brothers at the time, suspected that having permanent cover crops could be cost effective. "When fuel went from 25 cents to 75 cents per gallon," Garlock said, "I wondered why we were turning over all this dirt, spending time and fuel."

Garlock found that mowing existing vineyard weeds wasn't a success. When UC Davis representatives told him cover crops "didn't work," Garlock ignored their advice and sought assistance from Moore, with whom he had

Continued on page 93

Plants on a mission: Cover crop species used in Napa County

California brome
Bromus carinatis

California oniongrass
Melica californica

Meadow barley
Hordeum brachyantherum

Austrian winter pea
Pisum sativum

Restless natives working for the watershed

An evolving aesthetic and the numerous benefits of California grasses are propelling native perennials into the cover-cropping forefront. As many growers are interested in creating more natural watershed conditions in and around the vineyards, California native grasses are being selected for their heartiness, drought tolerance, filtration abilities and generally tough character. Native grasses can take a lot of abuse. Their relatively deep rooting system helps provide a wicking action that soaks up excess runoff and traps nutrients before they can enter the watershed.

Frequently used natives

Meadow barley   *Hordeum brachyantherum*
California barley   *Hordeum brachyantherum ssp. californicum*
"Molate" red fescue   *Festuca rubra, "Molate"*
Idaho fescue   *Festuca idahoensis*
California brome   *Bromus carinatis*
Blue wildrye   *Elymus glaucus*

Experimental natives

Pine bluegrass   *Poa secunda secunda*
various species of *Nassella*
various species of *Melica*, including
    California oniongrass   *Melica californica*
    and the Tory melic   *Melica torreyana*
California oat grass   *Danthonia californica*

Traditional no-till

For no-till situations, the traditional Mediterranean annuals are selected for their ability to provide erosion control and some other benefits. Because they establish more quickly than the native grasses —coming in right after a rain, rather than taking years to establish—they are more appropriate for urgent hillside erosion control.

Frequently used no-till

"Zorro" annual fescue   *Vulpia myruos*
"Blando" brome   *Bromus hordeaceus*
Annual rye grass   *Lolium multiflorum*
Red oats   *Avena sativa*
Cereal barley   *Hordeum vulgare*
Cereal rye   *Secale cereale*
Rose clover   *Trifolium hirtum*
Crimson clover   *Trifolium incarnatum*

Soil builders

Soil builders grow quickly and produce a lot of biomass. Tilled under, this biomass adds organic matter to the soil. The legumes in this mix enhance soil nitrogen. The quick germination of soil builders make them useful for erosion control in new vineyards—followed later by no-till crops (see above).

Frequently used soil builders

Red oats   *Avena byzantina*
Cereal barley   *Hordeum vulgare*
Cereal rye   *Secale cereale*
Bell (fava) bean   *Vicia faba*
Austrian winter pea   *Pisum sativum*
Vetch   *Vicia sativa*
Purple vetch   *Vicia benghalensis*
"Lana" woolypod vetch   *Vicia villosa ssp. dasycarpa*
Berseem clover   *Trifolium alexandrinum*
Crimson clover   *Trifolium incarnatum*

Case study: Going native, guarding soil
Lake Hennessey/Chiles Creek watershed

▲ Liesel and Volker Eisele

At Volker Eisele Family Estates, an organic farming operation in Chiles Valley, vineyard cover crops of perennial native grasses help conserve soil and water resources.

With his dog, Wolf, in the middle of a scorching day, Volker Eisele takes refuge in the shade of an oak beside the vineyard reservoir as he talks about his cover cropping practices. The first thing Eisele stresses is that he is an organic farmer. Since he lives on the land he farms, his motives for organic farming are rooted in pure pragmatism. "We haven't used poisons here in 27 years," he said, "because I knew the poisons would end up in our well."

He also knows that most of his healthy soil would end up in Chiles Creek if he didn't have cover crops.

Eisele takes for granted the need to consider water quality and erosion control as part of farming. "There isn't a single human being who doesn't live in a watershed," he said, explaining that erosion is a problem in every watershed, but is particularly serious in climates such as Napa County's, where the heat in summer "leaves the ground dry and pulverized" and "concentrated winter rains" create heightened conditions for erosion.

"Ideally one should have a solid cover crop on any given field," Eisele said, both for soil health and erosion control, but "erosion prevention is reason number one. If the soil is gone, soil health won't do me any good."

On the Eisele land, 60 acres have been planted to grapes. Trees and wildlife buffers control erosion on the remaining 310 acres of oak woodland and riparian vegetation, where volunteer trees are routinely protected as part of the land management practice.

In the vineyards, where most of the erosion has occurred on slopes under 3 percent, Eisele stopped disking in the 1980s as an experiment in erosion control. He let the weeds grow up naturally and mowed them, but later found that the vine prunings left on top of the weeds created ideal habitat for cane borers, and some of the natural weeds included noxious varieties such as yellow starthistle.

"So we went back to conventional tillage," Eisele said. The second experiment was with an erosion control

90   *Vineyards in the Watershed*

◀ Native grasses as cover crops help reduce vineyard erosion at the Eisele estate.

blend of grasses and clovers that his wife, Liesel, used in her landscape architecture business. The grass blend worked, but Eisele wanted native grasses.

"I've been what people would call a nut about native vegetation ever since I moved here," Eisele said, aware that non-native plants were a problem throughout the state. Eisele consulted with the NRCS, learning more about cover cropping. Then he struck up a friendship with Bob Bugg, a research scientist from UC Davis, whose knowledge of native grasses helped set the course for Eisele's cover cropping.

"He did something that really intrigued me," Eisele recalled. "We took a walk around the edges of the property, and he found remnants of native grasses. Native grasses are very difficult to identify."

Bugg suggested Eisele plant the grasses that were native to the property. "He encouraged me strongly to go native," Eisele said. "And so he figured out a mix for me and we tried it."

In 1996 the grass seeds were sown and the results were unusually positive. "Lo and behold," Eisele said, "the first planting we did was a resounding success. I've never had such a resounding success since." They planted Molate fescue, meadow barley, Idaho fescue and pine bluegrass.

◀ Wolf, the Eiseles' dog, leaves the reservoir after a cool drink.

Chapter Five: Cover Crops

Since that time there have been five plantings of cover crops—and some wildflowers thrown in for beauty's sake near the home. Challenges he has faced with cover cropping include weather. In an unusually cold autumn, Eisele had trouble with a cover crop planting because the ground was not warm enough for the seeds to germinate properly.

In recent plantings, meadow barley has been replaced by California barley, a less aggressive competitor.

In order to reduce the seed banks of invasive weeds before he plants his native grasses in the fall, Eisele uses overhead sprinklers to encourage weed germination, then disks the weeds. He repeats the process if he can squeeze in the time, then plants the cover crop. "On my hillsides where I don't have overhead sprinklers," he said, "all I can do is disk the hell out of the ground and pray."

Native grasses are not big drinkers. "One of the reasons I like native grass is that it goes dormant during the heat. That's part of its evolutionary development. It turns brown and barely lives," he said. But even native grasses do require a bit of water, so there is inevitable competition between grass and vine.

"I haven't solved the competition problem yet," Eisele said. But as he works on it, and even as erosion to some extent is inevitable, his native cover crops are doing their share to help keep soil on the property and out of Chiles Creek. ■

◀ Native grasses in the vineyard blend with the natural landscape.

Continued from page 88

worked on establishing hillside cover crops. In 1977 and 1978 Moore and Garlock set up Valley floor trials of no-till cover crops on a few half-acre plots.

Small plants, great achievements

Cover crops proved their worth in the first trials, saving operation costs and reducing soil compaction. At the Oakville test site, where poor quality fruit had historically resulted from vines with too much vigor, the cover crops did more than cut fuel costs. "Vegetative growth went down and yields went up," Garlock said. Then they noticed improved fruit character.

Garlock, who now works for Robert Mondavi Winery, said, "Out of those trials we found that when you reduced excessive vigor, you improved the quality of wines."

Later they also found that too much of a good thing was bad for vines, since excessively stressed vines were more susceptible to disease. But continuing experiments taught farmers how to manage the cover crops to the best advantage of the vine.

Other farming improvements came out of experiments with cover crops. First, where cover crops had been planted, there was greater control of dust and mites.

The critical need to feed soil communities

Many growers are still in the dark about soil microbial communities.

In the same teaspoon of healthy soil that may hold hundreds of millions—or even a billion—microorganisms, there are likely to be up to 5,000 different kinds of organisms, says Kate Scow, a soil microbial ecologist at UC Davis.

All plants critically depend upon the work of these organisms, which are often harmed by the wrong farming practices.

According to Scow, most soils are in a starved state, so the addition of cover crops is one of the best ways of adding carbon for the microbes. In turn, these unseen organisms provide critical services, including the breakdown of pollutants such as some pesticides that may be present in the soil. Breaking down contaminants, Scow explained, is a "well established benefit of active microbial communities."

They may also shut out pathogens. "There seems to be some support for the hypothesis," Scow said, "that there is resistance to pathogens associated with having a large soil microbial community."

Microbes build up organic matter and soil structure, which help in preventing soil loss. They recycle nutrients too, including the most essential plant nutrient, nitrogen.

"In fact," Scow says, "at every step in the complicated nitrogen cycling process, there is a group of microorganisms trying to make a living."

More discussion of soil biology and the critical functions of microbes can be found in "Soil," Appendix IV, page 164. ■

Chapter Five *Cover Crops*

▲ In the absence of cover crops, wet tractor roads can become slick and compacted. In dry weather, such roads can erode as dust, degrading air quality and reducing topsoil.

There was a reduction in soil compaction as well. Soil compaction can inhibit the soil's ability to exchange oxygen and carbon dioxide—basically preventing the organisms in the soil from breathing.

Soil compaction can also lead to erosion and flooding, even on flat ground, because compressed earth will not absorb water. Growers found that cover crops provided a cushiony surface and traction for equipment access during wet, spring months.

It has also become important to farmers that nitrogen-fixing cover crops can reduce the need for nitrogen fertilizers, which can travel in rainwater runoff and contaminate public water resources.

Learning curve

In the past three decades of trial and error, winegrowers have learned from experience to use cover crops to control erosion, manage vine vigor, adjust nutrients and organic matter in the soil, improve water filtration in the soil, increase vineyard biodiversity, attract and host beneficial insects, control weeds and—as the bare-earth aesthetic slowly dies—increase the beauty of the vineyard.

As cover crops began to change the aesthetic in the Valley, they also changed the thinking and the practices of growers who were willing to learn by trial and error. Those practices are continually evolving, since plants will respond differently to the variable conditions at each farming site.

While there is sufficient shared knowledge to help any grower using cover crops for the first time, some of those who helped pioneer the practice in the Napa Valley recall their early trials with a sense of humor. "I can't tell you how many bad cover crops I've grown to get to where I am now," confessed Zach Berkowitz, a long-time grower.

Berkowitz said that cover crops

◀ Six inches of "moon dust" once covered this tractor avenue between vineyard blocks at Schramsberg Vineyards until vineyard manager Craig Roemer planted the road with cover crops.

keep farming from ever getting boring because you get to try new things all the time. And when the crops didn't do exactly what they were planted to do, Berkowitz said he would humorously remind his vineyard crew, "We don't make mistakes, we just learn things." *(See page 100.)*

Underground benefits

Just as cover cropping moved, literally, from the hillsides to the Valley floor, its versatility moved attention from the surface task of erosion control to the underground service of improving the soil itself.

Kirk Grace, of Robert Sinskey Vineyards, appreciates how the waste products of cover crops feed those "unseen microbial populations" and support subterranean diversity. "In just a teaspoon of healthy soil there are up to 600 million organisms," Grace said. The more the better *(see page 93)*. Healthy communities of microscopic critters are what help cycle soil fertility—and they allow Grace to farm without using chemical fertilizers.

"By planting cover crops you have enriched the soil which, under a traditional paradigm, would be more sterile," said Grace. Adding synthetic products to soil may be fine in small quantities, he explained, but if overused they can burn soils, killing

◀ For better vine health, Kirk Grace, of Robert Sinskey Vineyards, avoids conventional chemical inputs and uses cover crops to build organic matter in the soil.

Chapter Five: Cover Crops

Chapter Five *Cover Crops*

▲ Rose clover is used as a cover crop.

organisms in the soil and creating a population shift that isn't necessarily a good one. "Where there's a vacant population space, Mother Nature will send colonizers to fill it," Grace explained. "A lot of times those pioneers can be pathogens."

For Grace, soil—its health, fertility and microbial activity—is a main building block of sustainability. "Our soil organic matter has gone from less than 1 percent 10 years ago to over 3 percent today," Grace said. And in addition to proper tillage and mowing techniques as well as natural soil amendments, Grace credits cover crops as an important tool used in achieving that success.

The little plant that could

Experiments with cover crops have been under way in Napa County for over 30 years. These experiments are still in progress, having so far demonstrated that cover cropping must be adapted on a site-by-site basis, taking into account the unique attributes of each vineyard.

In the process of learning the often mysterious ways of cover cropping, growers have used these plants—among other things—to control erosion, manage vigor, improve soil quality, increase organic matter in the soil, reduce soil compaction, filter nutrients and sediment, increase biodiversity, provide habitat for helpful insects, reduce weeds and create an alternative aesthetic for vineyards.

CHAPTER NOTES
[1] David Masumoto, *Epitaph for a Peach: Four Seasons on My Family Farm.* HarperCollins, 1995.

PHOTO CREDITS
Phill Blake: pages 86–87, 88, 89, 96
Juliane Poirier Locke: pages 90, 91, 92, 94, 95, 97 *bottom*
The de Leuze Family Vineyards collection: page 97 *top*

▲ This cover crop dwarfs the young vines planted at each protruding stake. This crop will be tilled under as "green manure" to build the organic matter in the soil.

◄ This vineyard is planted in alternating rows of tilled and non-tilled cover crops. This practice helps reduce the competition between vines and cover crops.

Chapter Five: Cover Crops

Chapter Six

▲ Merlot grapes await harvest. Well-managed vines can help prevent infestation of insects and disease.

Vineyards in the Watershed

Vine Health
Observing changes, integrating practices

Agricultural tradition dates back 3,000 years,[1] while modern chemical farming is only about 50 years old—the brainchild of chemical manufacturers who, after World War II, reformulated their chemical warfare materials for farm use.

Farmers were understandably drawn toward what appeared to be a silver-bullet solution to pest problems, and food crop production in particular became more efficient. But in time many people discovered that efficiency took a toll on the quality of life: Pesticides traveled in the environment, harmed wildlife, degraded natural resources, contaminated drinking water and endangered human health.

Continued on page 103

Improving conditions for vine health:
An interview with Zach Berkowitz

▲ Zach Berkowitz examines a Carneros cover crop.

Zach Berkowitz has been pursuing sustainable farming practices for decades. Among the first Napa County growers to experiment with cover crops, Berkowitz found over the years that improving vineyard conditions in a sustainable manner could help thwart disease and pests while improving vine health.

Zach Berkowitz, a vineyard consultant who for about 25 years managed the vineyards for Domain Chandon, believes that sustainable farming supports vine health and community good will. "Especially in a place like Napa, where you're farming around people all the time, be it neighbors, tourists or other businesses," said Berkowitz, "I think every farmer has to ask, how do I do it in a way that's most natural and not dependent on chemicals?"

Sustainable farming looks to natural, prevention-based care of vineyards, so that in the next century, winegrapes will still be grown in Napa Valley. The movement toward sustainability is a movement away from chemical-dependent farming practices, which are only short-term solutions. Berkowitz echoes the sentiments of many of his peers when he explains that chemical solutions put a grower on a treadmill, forced to treat a pest problem every year—which is not only expensive, but requires the additional hassles of training workers, wearing protective clothing, worrying the neighbors and becoming the target of criticism. "Let's be honest," said Berkowitz. "There's a lot of societal pressure against spraying chemicals."

Berkowitz understands why many farmers are reluctant to abandon conventional pesticides. "The beauty of the chemical way is you mix some stuff in a tank, you spray and you get results. So in the short term it's very attractive," he said. "But sustainability is about looking for long-term solutions."

The easiest first step for somebody who wants to grow more sustainably, Berkowitz advises, is cover crops. "In one step, you can potentially affect the whole range of vine health issues," Berkowitz said. Planting cover crops is among the sustainable strategies that address pest and disease conditions in a prevention mode. "In the triangle of disease, you need a host, a pathogen and the right conditions," said Berkowitz, adding that the pathogens and the hosts can't be changed, but conditions can.

Cover crops can improve vine health conditions—a preventative measure rather than an after-the-fact response to pests and disease already affecting the vine. "If you have big old vigorous vines and the fruit zone is shaded, it's difficult to get sulfur in and there's not much air movement through the vine. It's a condition more conducive to powdery mildew, which is a mold, and to botrytis, which is a rot," Berkowitz explained. "One thing people do is pull leaves, but I think everyone would agree that pulling leaves is a Band Aid activity, whereas if you could bring the vine into balance, you don't need the Band Aid anymore."

Vine balance refers to the optimum condition of the plant—somewhere between weak and overly vigorous. Planting cover crops can help a grower achieve vine balance by adjusting the competition between vine and cover crop. Berkowitz points out that such balancing is not a precise science. "Cover crops are tricky," says Berkowitz. "And that's why we keep trying to figure out how to make them work because each vineyard is a little different."

In 1983, at the advice of the Natural Resources Conservation Service (NRCS), Berkowitz planted a block at Domain Chandon with no-till cover crops before planting the pinot noir vines. Berkowitz told the NRCS he was worried about competition. "They promised it wouldn't happen," said Berkowitz. Unfortunately, it did. The vines turned yellow in June.

But after five years, those same vines produced more tons per acre than the adjacent cultivated vines. "Another thing we noticed," said Berkowitz, "was that when it got real hot, the no-till vines made it through better. The end of the story," he added, "is that the tilled section is now no-till."

The cost of seed mixes for cover crops compare favorably with conventional treatments, at a cost as low as $50 an acre, and the crops provide habitat for beneficial insects including predators to leafhoppers and other vine pests.

"All this sounds great, but it's harder than it sounds," Berkowitz said.

Sustainable farming looks to natural, prevention-based care of vineyards, so that in the next century, winegrapes will still be grown in Napa Valley.

▲ **Zach Berkowitz addresses attendees at a sustainable farming workshop.**

Chapter Six: Vine Health

▲ Cover crops provide habitat for beneficial insects and organic matter for soil organisms.

"It would be incorrect to suggest that all you have to do is buy a bag of seed and all your trouble will go away. There are no guarantees that if you plant a cover crop you won't have mites anymore. But you are improving the conditions for a more natural balance. The natural balance doesn't happen overnight, it happens with time."

After a few decades of figuring out what works at what site, Berkowitz likes to compare sustainable farming to a state of mind. "It's a new way of thinking," he said. "Once you realize the goal is this long-term thing, then you're more willing to explore and research."

For Berkowitz, sustainable farming requires continuous learning. "It's a trial and error thing. It's a patience thing," said Berkowitz. "There are a lot of successes out there, and we all can learn from them." ■

Continued from page 99

Sustainable farmers are moving away from high-risk pesticides toward more integrated pest control strategies. Many public interest groups and regulatory agencies support this movement,[2] and while a number of local growers have been developing a range of sustainable practices over many years in Napa County, others are increasingly interested in learning how to farm more sustainably.[3]

Growers are learning on a site-by-site basis what will work, and so the shift from conventional farming toward sustainable farming is gradual and slow.

IPM commandment: Know thy site

As farming returns to its centuries-old roots, and winegrowers re-evaluate the pest control challenges they faced long before chemical farming came along, an alternately effective and mysterious tool they are exploring is an approach known as integrated pest management, or IPM.

IPM cannot be strictly defined, but generally encompasses a complex and multifaceted approach to solving pest problems—one that includes prevention, environmental and biological controls and reduced chemical reliance. The complexity of IPM more closely mirrors nature, and the program encompasses pesticide use as a last resort.

Programs evolve in different ways, but the universal requirement is paying attention. As a starting point, a farmer takes stock of what's in and around the vineyard. In Napa County as elsewhere, the bugs, animals and diseases that show up in the vineyard correspond to an extent with the varieties planted, and can also say a lot about the conditions in the vineyard.

▲ The generous setbacks of this Mondavi vineyard help both the creek and the crop.

◄ RoundUp® is used to control weeds under these vines. Many growers are narrowing the width of the spray area to mitigate impacts and reduce costs.

Chapter Six *Vine Health*

Unlike the spray-first, ask-questions-later approach, IPM requires that farmers ask first whether and to what degree the grape variety being planted is susceptible to a pest of concern. The second question is, which practices might be used to prevent pest outbreaks?

A routine farming practice is fertilizing the soil. But when done as a rote task, this practice can contribute to pest problems. The over-use of nitrogen fertilizers has been linked to increased vigor and overly tender plant tissue in vines. This type of plant material is easier to snack on and makes an extraordinarily succulent feast for many pests.

Healthy vines can better resist pests.

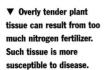

▼ Overly tender plant tissue can result from too much nitrogen fertilizer. Such tissue is more susceptible to disease.

The ever-threatening powdery mildew,[4] for example—for which the largest quantity of vineyard pesticide is used[5]—can be partially controlled by managing vine canopy; eutypa dieback[6] can be checked to an extent by pre-pruning vines, followed by hand-pruning vines as late as possible; botrytis bunch rot[7] can be prevented in part by leaf and shoot removal; and spider mites[8] can be discouraged by reducing dusty conditions (via cover crops, among other things). Where preventative measures have been taken, pest problems are likely to be less severe.

Together with preventative management practices, IPM looks to biological controls as a means of natural pest control. Because natural controls operate in a highly complex manner, they do not always work the way a farmer would like them to. As with any sustainable practice, trial and error are part of the process.

But progress is more likely when a farmer has a thorough knowledge of the vineyard site, its problems and potential problems. At the same time, growers need to learn about the pests and the natural predators of the pests, including the life cycles and habitat needs for each.

Beneficial insects and organisms

◀ Undeveloped land provides biodiversity that can benefit both vineyards and the community.

can be encouraged by the presence of natural habitat. Vineyards adjacent to natural vegetation, such as a healthy riparian corridor, have the advantage of greater biodiversity—that is, more kinds and numbers of organisms.

Of monocultures and biodiversity

Biodiversity addresses one of the most frequently heard public complaints about agriculture in general and viticulture in particular: the monoculture.

Ecologically speaking, a monoculture is neither natural nor sustainable. Extreme examples of monocultures exist at sites in California where all non-crop vegetation has been removed, in some cases for miles. In the state's lower Central Valley, for example, a person can drive for more than an hour and see only cotton crops. Such low diversity represents agriculture at its most chemical-dependent because the habitat for naturally occurring controls has been removed and the crop is thus more vulnerable to pest outbreak.

At the other end of the spectrum are highly biodiverse farms where many kinds of crops are grown together, incorporating natural vegetation. The organic Bon Terra Vineyards in Mendocino County incorporate insectaries in every other vine row, cover crops and vegetative land-breaks that include fruit trees and over 50 kinds of plants, including natives. The property covers 350 acres with only 132 planted to grapes, and two creeks provide natural riparian species. These high-biodiversity crop sites contain built-in biological controls, which help make the crops less vulnerable to pest outbreak.

Chapter Six *Vine Health*

▶ This native blend cover crop includes meadow barley, Molate red fescue, Idaho fescue and pine bluegrass.

Biodiversity in local vineyards

Napa County vineyards typically fall somewhere between the extremes of agricultural biodiversity. Some new vineyards have been developed without removing many—or any—trees. Some older Napa Valley sites were cleared of natural vegetation long ago and have been left unimproved. Other sites are being planted to increase biodiversity, often in conjunction with restoring riparian areas. Larkmead Vineyards in Calistoga, for example, is planting native vegetation along Selby Creek as part of a riparian restoration project.[9]

Many Napa County vineyards grow cover crops between vine rows. These cover crops provide numerous benefits including habitat for a diverse insect population, increased microbial activity in the soil and food for some organisms that might otherwise feed on the vines themselves. Where cover crops are grown along the peripheries of the vineyard, they can function specifically as insectaries, providing habitat for beneficial insects.[10]

Increasing the biodiversity in soil is another way of managing for pest control *(see sidebar page 93)*. Sustainable vineyards typically have a soil management program that can include composting, cover crops and erosion control.

▲ Preserved woodlands add biodiversity to this Hess Collection vineyard.

Managing for healthy, biodiverse soils

Avoiding pesticides is part of the management program at a number of sustainable properties, including the de Leuze Family Vineyards that produce some of the grapes used at ZD Wines in St. Helena.

The de Leuze family farms organically and used compost to improve the soil at a Carneros site, which was purchased as a vineyard "in decline," infested with phylloxera and producing gradually decreasing yields every season.

Owner Norman de Leuze said that when they purchased the failing vineyard and immediately converted it to organic in 1996, they "used an enormous amount of compost" to restore the soil, which was low in both phosphorus and nitrogen.

They applied compost for the first time at almost 36 tons[11] per acre (they now use about 4 tons per acre). The following harvest, yields increased by almost 2 tons per acre. Even after seven acres of the property were replanted in a low-yielding variety, the vineyard—which still has phylloxera and rootstock[12] that is not inherently resistant to the pest—is maintaining good yields and has about 3 percent organic matter in the soil.

"The tilth of the soil," de Leuze said, "is substantially improved." When he

▲ Norman de Leuze digs into some home-made compost.

▲ A compost pile awaits a vineyard application. Compost can help restore soil health and eliminate the use of synthetic nitrogen fertilizers.

Chapter Six: Vine Health

Case Study: Riparian management and Pierce's disease
Conn Creek/Napa River watershed

▲ Drew Johnson beside Beringer's Conn Creek restoration site near Yountville.

When Drew Johnson got his degree from UC Davis and came to work for Beringer Vineyards in 1989, he expected to be growing vines, not hardwoods. But as part of an experiment in natural pest control, he learned quickly about how trees and other creek vegetation could affect the amount of Pierce's disease that got to his vines.

Pierce's disease is carried by insects including the infamous glassy-winged sharpshooter, which is not present here but has done so much damage elsewhere that Napa County farmers have nightmares about the insect. But it's not the bug itself that threatens Napa County's core industry. "The glassy-winged sharpshooter won't kill any vines," one local farmer said. "It's the Pierce's disease that will. Pierce's disease is a real test for sustainable agriculture."

To date, the glassy-winged sharpshooter is not a local problem. But the smaller, blue-green sharpshooter has been established here for some years. It lives along waterways throughout Napa County, breeding in plants along the creek banks and carrying disease to grape vines.

Some riparian plants—mostly non-natives such as the Himalayan blackberry—act as hosts to the bacterium that causes Pierce's disease—Xylella fastidiosa. This bacterium lives in a plant's water-conducting system.

The sharpshooter is one of the few insects that can feed on the water-conducting part of the plant, take up the bacteria via specialized mouthparts and distribute it to the other plants they feed upon. When Pierce's disease infects vineyards, the vines become gradually unable to take up water.

At Conn Creek, near the Beringer property, the blue-green sharpshooter had infected the vineyard with Pierce's disease and, in the mid-1990s, Johnson participated in a unique approach to lowering the vector's populations.

When one landowner was cited and fined by regulatory agencies for bulldozing vegetation infected with Pierce's disease, the Natural Resources Conservation Service (NRCS) organized a

◀ Branches pruned from the hardwoods along Conn Creek.

think tank of experts to devise solutions rather than mere regulatory actions. Calling themselves the "Not the Silver Bullet Committee," the group included the NRCS, UC Berkeley, the UC Cooperative Extension, the state Department of Fish and Game (DFG) and Beringer Vineyards.

The committee's approach was to do a field study of riparian vegetation for Pierce's disease control. As part of the plan, Johnson oversaw the removal of host plants and revegetation with native plants that did not host vectors of Pierce's disease.

All the work was done with chainsaws, weed whips and hand labor. No heavy equipment was used. All the blackberries, wild grape, vinca, mugwort and small willows were removed.[18] Hardwoods were planted including oak, ash, maple, bay and walnut. "There was immediately a dramatic decrease in the number of blue-green sharpshooters in that area," said Johnson. "They probably decreased to about 2 percent of their original level."

Researchers found that wildlife and water quality were adequately maintained, and some staff of the DFG were relieved that there was now a means of addressing Pierce's disease that was good for the creeks. "The DFG didn't want to see people going into all the creeks and waterways and start gutting (them) because they had Pierce's disease," said Johnson. "That would have been really, really bad. But at the same time, we had to do something about Pierce's disease."

The study served as a basis for the county's task force recommendations for controlling Pierce's disease.

The extra work and the roughly $20,000 in costs have been well worth it to Johnson. He continues to manage the creek vegetation and plans to add some native ground cover to the creek soon, perhaps spice bush and native grasses. "It's a never-ending project," said Johnson. ■

"There was immediately a dramatic decrease in the number of blue-green sharpshooters in that area. They probably decreased to about 2 percent of their original level."

—Drew Johnson

Chapter Six: Vine Health

▲ Heavy applications of compost were used to improve the soil at this Carneros vineyard.

compares the soil of his vineyard to the soil of another Carneros vineyard—a vineyard routinely sprayed with pesticides—he says the difference is remarkable. "Our soil has a real nice, loose look," de Leuze said. "And his (soil) gets a hard, cracked look."[13]

While even those pesticides considered to be low-risk products can cause imbalances in the soil microbial populations, building the organic matter in the soil increases soil biodiversity and stability while also making nutrients more readily available for plant uptake.

Monitoring—observing the clues

In another grape-growing region of California, growers recognize the importance of monitoring vineyards. Those participating in a self-evaluation program gain program points for simply getting out of their trucks while visiting the vineyard. This is to get them closer to the ground where they can see—and ideally monitor routinely—the vines and the vineyard floor.

While there is no such point program in Napa County, growers using the IPM approach walk the vineyard to monitor a number of vineyard conditions that relate directly to pest outbreaks. These conditions can include soil moisture, pest populations, beneficial insect populations, vine conditions and others. Some growers use sticky traps to help monitor insect populations.

Kirk Grace, of Robert Sinskey Vineyards, explains that both random and routine monitoring is necessary for vineyard assessment. "Random checks you make once a week to look for bugs, disease or other changes," said Grace, who explained that routine checks are equally important. "Last year's spot for over-wintering mildew," he said, "would be a hot-spot this year, and should be routinely monitored."

These random and routine checks, according to Grace, require an "ability to see change and to interpret that change." In interpreting data, timing needs to be correct. For example, leafhoppers are present throughout the year, and a farmer may observe that a leafhopper population is going up in the vineyard. The farmer is then faced with a decision whether or not to react.

Don't panic

"The true mastery," Grace explained, "is seeing (the leafhopper populations) go up and not spraying because the population hasn't reached a critical mass. Populations go up, but the craft is being able to interpret when the economic threshold is going to happen."

The economic threshold[14] is the point at which some response is judged necessary in order to protect the vines. Determining this threshold is closely tied to an individual grower's tolerance and patience. Growers who spray at the least provocation—or worse yet, on a rigid schedule that does not correspond with actual vineyard conditions—are not farming sustainably.

"(Farmers who) like pesticides," says Glenn McGourty, of the Mendocino County UC Extension, like them because they are "very predictable in the outcome." Biological controls, on the other hand, are not always predictable. McGourty, who has done extensive work with cover crops in IPM, says that

▲ Top: Young vines and new growth are most vulnerable to pests.

Bottom: This vineyard in Carneros is free of synthetic chemical pesticides and fertilizers.

Chapter Six: Vine Health

Chapter Six *Vine Health*

▶ Barn owls, accustomed to living near human activity in barns, will readily nest in owl boxes installed near vineyards.

in Mendocino County there are numerous biological buffers (such as riparian areas) that work with cover crops to support a good pest balance in the vineyards, without pesticides. "What the growers up here in Mendocino County have found," said McGourty, "is that when they lay off the chemicals, things tend to fall in place."

Build it and they will come

Among the many pest-control services provided by nature are those performed by birds of prey. Across the road from Bothe State Park in the northern Napa Valley, owls that breed in the park's trees forage in the vineyards that lie just across Highway 29.

Not every vineyard has a state park so close by, but many properties have owls living in adjacent woodlands that were left undeveloped when the vineyard was put in. On sites where trees are scarce, many growers can lure helpful bird and bat species to live and hunt in vineyards.

Although bird boxes are not well studied yet in a vineyard landscape[15] and in hot climates birds can become overheated in certain box designs, many farmers have employed safe nesting boxes to draw pest-eating birds and bats to the vineyard.

▲ A bat colony can eat up to five pounds of insects per night.

Some Napa County growers are participating in the Bluebird Recovery Program of the California Audubon Society, installing bird boxes for the native western bluebird, which feeds on vineyard insects.

Monitoring year round and cleaning the boxes after nesting is completed are important to a vineyard management plan that includes bird boxes. In the course of monitoring boxes, a grower can discourage any European starling that tries to use the nesting box.

Nesting boxes for owls are popular in local vineyards, since a barn owl will nest in a box, then forage for and eat hundreds of vineyard rodents while raising a family. Mice, voles and gophers are favored food for owls. The foraging habits of barn owls can be especially helpful where vole and gopher populations are using cover crops for habitat.

Social sustainability can be promoted by using owl boxes as well. Local non-profit groups benefit from owl box sales through a countywide program called Habitat for Hooters.[16]

Many bat species are voracious insectivores, and some species can also be lured into the vineyard with bat boxes. It's trickier to keep bat populations in the vineyard, but possible for anyone willing to put forth the effort.[17] Bats eat numerous insects including moths, cucumber beetles and leafhoppers. Entire bat colonies can eat five pounds of insects each night.

Small steps, more learning

Drawing beneficial birds or bats into vineyards does not entirely solve pest problems. But as part of a pest control program that integrates preventative practices with biological controls and a watershed farming approach, the presence of bats, barn owls and bluebirds can be understood as part of the bigger effort—to increase biodiversity on the farm and to continue learning all the ways in which natural controls can help improve the health and balance of the vines and the environment in which they grow.

In the progression toward sustainability, bat, owl and bird boxes are replaced by trees and other natural nesting places.

▲ **Top:** This owl box is placed outside the vineyard so farming practices will not disturb the birds. The perch has been removed from this box because barn owls will not use the perch but their predators will.

Below: Bird netting is an alternative to the conventional practice of trapping and gassing birds. It offers full protection of grapes without harming wildlife.

Chapter Six: Vine Health

Pesticides

Vineyard pesticides used in Napa County[1] 2000

Pesticide	Pounds applied	Acres treated[2]	Potential groundwater threat[3]	FQPA research priority[4]	PANNA rating[5]	Wildlife toxicity rating[6]
Fungicide *(kills fungus)*						
Azoxystrobin *Abound®*	1,910	10,209	-----	Priority II		-----
Benomyl *Benlate®*	1,570	3,925	-----	Priority I	"Bad actor"	Highly toxic to fish
Cyprodinil *Vangard®*	1,527	3,738	-----	Priority II	"Bad actor"	-----
Dicloran *Botran®*	411	289	Yes	Priority II		-----
Fenarimol *Rubigan®*	361	10,170	----	Priority II		-----
Fenhexamid *Elevate®*	289	579	-----			-----
Iprodione *Rovral®*	2,169	3,381	Yes			Moderately toxic to fish
Lime-sulfur	2,874	684	-----			-----
Mancozeb *Dithane®*	3,775	3,392	-----	Priority I	"Bad actor"	Highly toxic to fish
Myclobutanil *Rally®*	2,472	29,210	-----	Priority I		Moderately toxic to fish, birds and mammals
Potassium bicarbonate *Kaligreen®*	10,039	4,187	-----	Priority I		-----

Vineyard pesticides used in Napa County[1] 2000

Pesticide	Pounds applied	Acres treated[2]	Potential groundwater threat[3]	FQPA research priority[4]	PANNA rating[5]	Wildlife toxicity rating[6]
Fungicide *(kills fungus) continued*						
Propionic Acid	267	2,739	-----			-----
Sulfur	1,850,979	216,763	-----	Priority II		Practically non-toxic to fish Slightly toxic to birds, mammals
Tebuconazole *Elite45®*	948	9,458	-----			-----
Trifloxystrobin *Flint®*	368	3,032	-----			-----
Triflumizole *Procure®*	839	5,463	Yes	Priority I		-----
Biocide *(kills everything)*						
Methyl bromide	12,039	34	-----	Priority II	"Bad actor"	Moderately toxic to wildlife
Chloropicrin	10	6	Yes		"Bad actor"	-----
Insecticide *(kills insects)*						
Carbaryl *Sevin®*	107	128	Yes	Priority I		Slightly to moderately toxic to birds, mammals, fish
Cinnamaldehyde *Valero®, Cinnacur®*	612	184	-----	Priority II		-----
Neem oil extract	118	180	-----	Priority II		-----
Cryolite *Kryocide®*	315	54	-----			Slightly toxic to fish and wildlife

Chapter Six: Vine Health

Pesticides

Vineyard pesticides used in Napa County[1] 2000

Pesticide	Pounds applied	Acres treated[2]	Potential groundwater threat[3]	FQPA research priority[4]	PANNA rating[5]	Wildlife toxicity rating[6]
Insecticide *(kills insects) continued*						
Diazinon *Knox-out®* *DiazinonAG500®*	2,208	37,869	Yes	Priority I		Extremely toxic to birds and mammals
Dimethoate	393	1,378	Yes	Priority I		Avoid wildlife exposure
Fenamiphos *Nemacur®*	138	58	Yes	Priority I		Extremely toxic to fish and wildlife
Imidacloprid *Admire®, Provado®*	506	11,471	-----	Priority II		Extremely toxic to fish and wildlife
Acaracide *(kills mites)*						
Fenbutatin oxide *Vendex®*	302	337	-----	Priority II		Highly toxic to fish / Extremely toxic to wildlife
Propargite *Omite®*	40	26	-----	Priority I	"Bad actor"	Highly toxic to fish / Lightly toxic to wildlife
Rodenticide *(kills rodents)*						
Aluminum phosphide *Phostoxin®*	14	28	-----			Extremely toxic to wildlife
Chlorophacinone	Not listed	154	-----			Extremely toxic to wildlife
Diphacinone	.03	968	-----			Extremely toxic to wildlife

[1] This list includes vineyard pesticides most commonly used in Napa County.

[2] This number includes multiple treatments. For example, if one acre was treated 10 times it will be listed as 10 acres treated.

[3] Source: California Department of Pesticide Regulation (DPR) and staff biologists for Napa County Agricultural Commissioner's office.

[4] Under the Food Quality Protection Act of 1996, pesticides will be phased out according to priority status assigned to each chemical product.

[5] Source: *Hooked on Poison: Pesticide Use in California 1991-1998*, by Pesticide Action Network North America. San Francisco: Californians for Pesticide Reform, 2000. PANNA is a public watchdog organization concerned with reducing pesticide use and promoting sustainable agriculture.

[6] Source: California Department of Fish and Game (DFG) report, *California Wildlife and Pesticides*.

CHAPTER NOTES

[1] J.J. Meisinger et al. "Effects of Cover Crops on Groundwater Quality," in *Cover Crops for Clean Water,* W.L. Hargrove, ed. Soil and Water Conservation Society Conference Proceedings, 1991: pp. 57-68.

[2] The Environmental Protection Agency (EPA), the California Department of Pesticide Regulations (DPR), the Napa County Agricultural Commissioner, Pesticide Action Network North America (PANNA), Physicians for Social Responsibility (PSR), World Wildlife Fund (WWF) and the Natural Resources Defense Council (NRDC), to name a few.

[3] The Napa Sustainable Winegrowing Group (NSWG) workshops for watershed-based farming practices have been well attended over the past five years.

[4] *Uncinula necator*

[5] More sulfur is applied to California vineyards than any other pesticide. Used to prevent powdery mildew, sulfur is a low-risk pesticide; however, the rate at which it is applied, together with spray techniques, can result in drift, causing air-quality degradation and health problems for those who have respiratory problems and allergic reactions to sulfur. The California Winegrape Pest Management Alliance is educating farmers to improve sulfur application.

[6] Caused by *Eutypa lata* fungal pathogen.

[7] *Botrytis cinera,* often called botrytis bunch rot.

[8] *Tetranychus pacificus* and *Eotetranycus williamette,* plant-feeding mites.

[9] A partnership project conducted by the Resource Conservation District (RCD) and the Natural Resources Conservation Service (NRCS).

[10] The benefits (and some drawbacks) of cover cropping are discussed throughout this book and highlighted in Chapters 1, 2 and 3.

[11] This weight is calculated by converting 3,000 yards at a rate of 3 yards per ton.

[12] AXR rootstock is a hybrid that does not present the genetic resistance to phylloxera.

▲ Juvenile barn owl

[13] Unhealthy, compacted soil can appear similar to concrete when dry. Visual and other soil assessment techniques can be found in Appendix IV.

[14] Since there is seldom economic data available upon which to make a field decision, one industry observer suggests that "action threshold" might be a more accurate term than "economic threshold."

[15] According to Adina Merenlender, UC Research Station, Hopland.

[16] For resources on bats and bat housing, see Appendix V.

[17] Mugwort and willow are important native plants. Removing these plants is a benefit to vineyard management, but not necessarily a benefit to the riparian habitat, according to UC Davis researcher and NSWG member Bob Bugg.

PHOTO CREDITS
Kathy Kellerson: pages 98–99
Juliane Poirier Locke: pages 100, 102, 103, 104, 105, 106 *top*, 108, 109, 117
Astrid Bock-Foster: page 101
Richard Camera: page 106 *bottom*, 107 *bottom*, 111 *top*, 113
Collection of de Leuze Family Vineyards: page 107 *top*, 110, 111
William E. Gill: page 112
Photo by Dr. Merlin D. Tuttle, courtesy of Bat Conservation International, Inc.: page 112 *bottom*

Chapter Six: Vine Health

Chapter Seven

▲ How vulnerable young vines are protected often defines the boundaries of a sustainable farming program.

Vineyards in the Watershed

Organic Farming
Profitability and land ethics

Before synthetic chemical pesticides came to the Napa Valley, every grower farmed organically. Louise Rossi, a 93-year-old St. Helena winegrower—who recalls horse-drawn tillage and the buried pipeline that once sent wine sloshing its way underground, straight from her father's winery into the Southern Pacific train cars—was not opposed to synthetic farming chemicals when they came along.

Rossi, who still lives in the family farmhouse surrounded by vineyards, grows her winegrapes organically now and claims she doesn't need the synthetic pesticides used in mainstream farming. "There aren't any bugs out there now," Rossi explained rhetorically. More precisely, the insect populations in her organic vineyards are apparently in balance.

Continued on page 122

Case Study: Organics and the bottom line
Napa River watershed

At the Yount Mill Vineyards, where family-rooted business and land ethics span four generations, profit and stewardship are tandem goals.

▲ Mustard is a cover crop at Yount Mill Vineyards.

Andy Hoxsey is no tree hugger. At the Napa Wine Company in Yountville, the slogan posted over his desk reads, "Happiness is positive cash flow," and the only nature photos in the room are shots of a family cattle ranch.

A fourth-generation grower who bristles at the mention of regulation, Hoxsey seems an unlikely poster-child for earth-friendly farming, but his 600 acres of winegrapes are organically grown. For Hoxsey and vineyard manager Jim Del Bondio, organic farming is good business and no extra trouble.

"It's like falling off a log," said Hoxsey, who calculates that his cost-per-acre for organic farming is the same as what he formerly paid for conventional farming. "It's almost easier than growing conventionally."

Hoxsey began transitioning Yount Mill Vineyards from conventional to organic almost 10 years ago, in part because he suspected that increasing regulations would further encroach upon chemical farming options anyway, and in part because of a personal land ethic passed down through generations.

"I remember my grandfather saying that nobody actually owns the land. You're stewards," Hoxsey said. "That always meant something to me." Del Bondio—who first ran a tractor on the property at age 7, and whose own organic vineyards total 50 acres that have been family farmed since the early 1900s—added another reason for the shift in practices. "We feel that the employees benefit from organic farming," Del Bondio said.

Hoxsey and Del Bondio worked with organic farm adviser Amigo Bob Cantisano to transition the Oakville and Yountville plots to organic, beginning in 1993. Hoxsey had to study his own soils in the process of learning how to farm organically. "In the early days I was scratching my head," said Hoxsey. There were areas in his vineyard where compost was not breaking down, and other areas where organic matter was accumulating; soil tests showed a range of 1.5 to 6 percent organic matter. "Our organic matter was going the wrong way in both directions." Cover crops weren't doing well.

Amending the soil with natural inputs achieved soil balance. Now the soils average 4 to 4.5 percent organic matter throughout, largely because of composting and cover crops. "In the early days, we were mining our soils,"

Hoxsey explained, which led them to try commercial composts. They later found that homemade composts made the most remarkable difference in soil balance. According to Hoxsey, both winegrape quality and yields have improved as a result of devising the right compost recipe, including no more than 20 percent each of pomace, horse bedding straw, turkey bedding straw, wood chips and pre-composted green waste.

"Our cover crops are stunning," Hoxsey points out, explaining that—in contrast to nitrogen fertilizers—plants make nitrogen available to the vines in such a way that it won't leach into groundwater. Hoxsey noted that the grains, peas and mustard crops planted between vine rows also help control erosion and dust as they simultaneously build up soil health.

"I'm not a zealot," Hoxsey said. "I'm not going to chide the next farmer and say, 'You've got to do this.'" He does it because he believes it's the best stewardship he can provide while maintaining that positive cash flow.

"I'm not going to do the community, my employees, my family or myself any good if I'm not here next year," said Hoxsey. "I do the community much better stead by being profitable." ■

▲ Andy Hoxsey, at Yount Mill Vineyards in Oakville, works with Jim Del Bondio to maintain 600 acres of organic winegrapes.

"I'm not going to do… my employees, my family or myself any good if I'm not here next year. I do the community much better stead by being profitable."
—Andy Hoxsey

Chapter Seven *Organic Farming*

▲ The Rossi vineyard in St. Helena is organically grown.

▶ Louise Rossi, organic winegrower.

Continued from page 119

The relatively low pest density in Napa County might partially explain why many of Rossi's contemporaries continued without the new chemical products, and why regional use of synthetic pesticides did not mirror pesticide use in other parts of the country.[1] Nevertheless, most growers in the area use synthetic chemicals, even in some sustainable vineyards.

How sustainable relates to organic

To achieve sustainability, both the material being applied to the vineyard and the means of applying it must be taken into account.

Among synthetic and non-synthetic farming chemicals, those approved by the Organic Materials Review Institute (OMRI) for organic farming are thought to pose the least threat to the health and safety of animals, humans, water quality and the environment. But if a worker must make several passes with a tractor to apply the approved-for-organic substance, the hidden costs of multiple tractor runs include air quality degradation, extra labor hours, fuel costs and soil compaction.

On the other hand, a synthetic chemical might be applied with a single tractor run—therefore saving on fuel and labor costs, air quality degradation and soil compaction—but if the chemical is highly toxic, then the human health and water quality degradation must also be weighed in order to determine sustainability.

Those who are attempting to farm sustainably use a systems approach. They examine the big picture before determining what is most sustainable, whether they are using fewer synthetic chemicals in less volume and applying them with greater safety and precision, or using only those crop applications allowed in organic growing.

Some vineyards in Napa County have been grown organically for generations without certification. Others—some of which were formerly certified organic—mix organic practices together with other techniques to create a sustainable operation.

Growing like crazy

No one knows how many farm acres in California are grown organically, because not all of them are registered with the state. Many people farm organically without any certification or with a form of self-certification. Others farm organically with third-party certification from one of more than 12 organizations. Even the most thorough tally of organic acreage is based on records that are two years old.

"Whatever number of (certified organic) acres you get will be outdated tomorrow," said Ray Green, the organic program manager with the California Department of Food and Agriculture. According to Green, the number of organic farm acres in the state is "growing like crazy."

Sulfur—improving what comes naturally

The one substance used in both organic and non-organic viticulture is sulfur, applied to prevent powdery mildew on grapes. Although sulfur is a product of nature, the stuff growers use is synthetically produced from three industrial processes: recovered from coal combustion, steam extracted from natural gas or refined from petroleum.

Until just over a decade ago, organic growers had to use only naturally occurring, mined sulfur. However, since mined sulfur was found to have impurities including arsenic, which is also naturally occurring, synthetically produced sulfur was approved for organic farming practices. Used on both organic and non-organic vineyards, synthetically produced sulfur is the more pure of the two kinds.*

While synthetic sulfur is relatively pure, the sulfur products used on vineyards contain additives and inert ingredients not discussed here. ■

In Napa County, 20 winegrape vineyards—totaling 1,146 acres—have been organically certified by the California Certified Organic Farmers (CCOF), a process that begins with an application and culminates three years after any prohibited materials have been used on the vineyard. The CCOF is the largest of the certifying bodies in California, and statewide it has certified 5,224 acres of organic vineyards.

Organic growers must submit to at least one inspection per year and ongoing record-keeping requirements. "The record keeping required to be certified is really no different than the record keeping that's required to run a

Continued on page 126

Perspectives on organic farming: Pauline and Vince Tofanelli
Napa River/Simmons Creek watershed

▲ Pauline Tofanelli

▲ Vince Tofanelli

The Tofanellis are old-time Valley farmers who grow wine grapes organically.

When asked why she has never used synthetic chemicals on her vineyards, Pauline Tofanelli, whose home adjoins the Tofanelli Vineyards in Calistoga, responded with a good-humored shrug and a smile. "Stubborness, I guess," she said. "We were used to using our hands and our backs."

Wearing work gloves and a straw hat, 76-year-old Pauline took a break under a shade tree with her son, Vince Tofanelli, on a June day so hot the only visible activity was a Plymouth Rock hen stepping madly across the yard toward the garden.

The mother-and-son farming team talked about the family's Valley floor property, farmed since 1929 and transitioned from pears and prunes to winegrapes beginning about 1942. The 26 acres of dry-farmed vineyards bordering Simmons Creek include three acres with vines over 70 years old.

The Zinfandel, Charbono, Sauvignon Blanc and Semillon winegrapes at Tofanelli Vineyards are grown without the chemicals used in conventional farming.

While they may go through the process of organic certification in the future—at the request of winemakers —the Tofanellis haven't yet bothered.

"We don't need to prove anything. We just continued farming the old way," Vince said. "I felt it was a sound way of farming."

Vince, who began farming in the early 1970s, explained that his decision to farm organically was in part influenced by the thinking of the times, when "environmental awareness was in the forefront."

He did experiment with conventional pesticides once about 20 years ago when, during a bad infestation of mites, someone suggested he spray. He was unhappy with the side effects in the vineyards.

"I killed off the good bugs, too," Vince explained. "The following year we decided to just bite the bullet." He hasn't used the mainstream farming chemicals since, and losses from pests are now accepted as part of the cost of farming naturally.

The mites are still in the vineyard, but the good bugs are now back. "The last couple of years it seems there's more ladybugs around," Pauline said. ■

Perspectives on organic farming: Boz Scaggs
Napa River/Dry Creek watershed

Winegrowing "newcomer" Boz Scaggs believes in organic farming.

I feel like I've just begun to open myself up to the concept of sustainable farming," said celebrity recording artist Boz Scaggs. Slouching comfortably on a porch chair at his modest hillside home—catching the ridge top breezes and the Valley views below—Scaggs said, "I'm still learning."

Seven years in Napa County have given him an insider's perspective of the area. "The Napa Valley experience to some people suggests a luxury life, but when you get down to the reality of this place, it's an agricultural community," Scaggs said. "Everyone up here is connected to agriculture in one way or another."

His own connection to farming is held through two small, organically grown vineyards.

Even in the midst of completing his 13th recording, Scaggs keeps one eye on his winegrapes and feels a responsibility to farm what he considers the "right" way.

Smiling, he confessed his discovery that farming is "not all sunshine and dirt." Trying to take in what he calls "the science behind it all" feels overwhelming to him at times. "It's like drinking from a fire hydrant," he said.

Farming organically was an easy decision for him and his wife, Dominique. "We had an awareness we brought with us," Scaggs explained. "We never intended to use any chemicals." The couple knew they would manage the landscaping, vegetable garden and vineyards organically when they considered what they "wanted to live near and to breathe."

They live close enough. You could throw a rock from the porch and hit the vineyard, where a madrone tree stands in the middle of a vineyard block. "We took out very few trees," Scaggs said.

On the 25-acre hillside property, only 1.5 acres were converted to vineyard, and the Mourvedre, Grenache and Syrah will become a southern Rhone style of wine.

There are no plans to expand. Scaggs and his wife are satisfied with less than two tons of winegrapes per acre. "I've got more than enough for our needs," Scaggs said. ■

▲ Boz Scaggs, recording artist and part-time winegrower.

▼ Part of the Scaggs' organic vineyard, where most trees were retained in the vineyard development.

Chapter Seven: Organic Farming

Chapter Seven *Organic Farming*

> "Hess was determined to be organic, but unfortunately it didn't work. I think that establishing young vines using RoundUp® and then switching to a non-chemical weed control is a sustainable practice."
>
> —Richard Camera,
> The Hess Collection Winery

Continued from page 123

business," said Brian McElroy, CCOF director of grower certification, "and to know how your resources are allocated."

Sustaining a mix of practices

Some vineyard managers have decided that following organic regulations is not always the most sustainable route for a vineyard. The 70-acre Veeder Summit Ranch, owned by The Hess Collection Winery, had just been certified organic in 1992 when workers discovered phylloxera. About half the ranch was newly established plantings under a lot of stress. After years of coaxing, new vines weren't getting up the stakes.

While a full-time, four-man crew used weed-whips on star thistle, the thistle's root systems were competing with new vines. Compost wasn't helping the vines because it wasn't getting to the vine roots and couldn't be tilled under. Liquid fertilizer couldn't be added to the drip system because at that time it wasn't approved for organic vineyards, and the frequent watering needed to boost the young vines seemed to be only promoting the growth of the weeds.

So when phylloxera was discovered, it was decided that the young vines might not be hearty enough to survive unless the star thistle was killed. They used RoundUp® on about 35 acres of vines. The vineyards were saved while organic certification was lost.

"Hess was determined to be organic, but unfortunately it didn't work," said Richard Camera, vineyard manager who came on board just after the crisis. Camera is experimenting with a weed badger *(see photo page 68)* to achieve affordable, non-chemical weed control on hillside vineyards, and believes that on new vines RoundUp® is still needed for the first few years. "I think that establishing young vines using RoundUp® and then switching to a non-chemical weed control is a sustainable practice," Camera said.

Up close and personal

Others won't consider even small applications of synthetic herbicide as part of their sustainable program. Daphne Araujo, of Araujo Estate Wines in Calistoga, explained that conventional chemicals are "not the way organic growers want to go."

At their Eisele Vineyards—a property sustaining winegrapes since the 1880s—Daphne and her husband, Bart, have recently transitioned their 41 organic vineyard acres to biodynamic

▲ The garden-like landscaping at Araujo Estate Wines in Calistoga unifies the vineyards, groves and gardens—all organically grown.

(see page 128). The Araujos recognize that the relatively small size of their vineyards—less than a quarter of the total property size—affords them the opportunities to do things that might not be sustainable on a larger scale.

Weed control, for example, is done by hand and by hoe plow. Because their estate wines can command a premium price, the labor costs are sustainable. Regardless, the Araujos are "always looking for ways to bring the costs down," while remaining committed to a land ethic that eschews the use of synthetic chemicals.

For Araujo, any measure of farming sustainability includes not only the health of the farm but of the farmer as well. When the "holistic remedies" used for biodynamic farming have been prepared and applied to the vineyard, not only has the vineyard become more resistant to insects, disease and sunburn, Araujo says, but the atmosphere she senses in the vineyard and among the workers seems exceptionally peaceful. This subjective experience is part of why she farms organically.

"Farming is about being in touch with the plants and the land," says Araujo. "We take our stewardship very seriously. We're only here for a relatively brief time, and then the land belongs to someone else."

Chapter Seven: Organic Farming

Case Study: Biodynamics—pushing the organic farming envelope
Napa River watershed

▲ Bulmaro Montes

▼ Ground crystal is used to make this horn silica, the basis of a preparation known as "501."

Viticulturalists for Joseph Phelps Vineyards are applying biodynamic farming practices to their vineyards—hoping to harness cosmic energy to benefit the soil, the vines and the wine.

Bulmaro Montes' father, like other farmers in Mexico, used to tie a red cloth in the branches of a tree to protect blossoms during eclipses, and place a horn in the crook of a tree to protect crops from insects.

Montes, in contrast, got his farming education locally, rising from field worker to director of vineyard operations for Joseph Phelps Vineyards. He has been growing winegrapes for 30 years.

Recently, however, Montes has taken up the practice of burying manure-packed cow horns at a certain phase of the sun and digging them up at yet another phase. Montes may be getting back to his roots in a way, but he is not reclaiming his father's farming rituals. What might appear as a fallback to ancient Mexican agriculture is actually part of a 20th century European practice known as biodynamics.

Austrian Rudolph Steiner, in the early 1920s, wove together biological and metaphysical principles in a holistic farming theory that he referred to as "spiritual science," which views the earth as an organism and purports that plants benefit from a special form of energy stored in "preparations."

A preparation known as "500" is made from the manure of female cows that has been buried in the horns of female cows after the fall equinox. When the spring equinox has passed, the horns are dug up and the manure is stored in a peat moss-lined box until it is mixed in water, stirred—so that the liquid creates vortex and reverse-vortex patterns—and applied to the vineyard.

Montes and viticulturist Philippe Pessereau are both learning to farm biodynamically. "It's like organic, only more complex," Montes explained. Part of the difficulty is understanding Steiner's complex translated works, and then adapting them so they are practical. Steiner advises, for example,

▲ Left to right: Constantino "Tino" Corro, Bulmaro Montes, Philippe Pessereau and Javier Nino de Rivera, the biodynamics team at Phelps.

hand stirring the preparations.

"You have to use common sense," said Pessereau. "In the beginning we were stirring the preparations by hand, but when you have 300 gallons, you need machines." When the preparation has been completed—which takes an hour per batch—it is applied to the vineyard at one-quarter cup of so-called "horn manure" to 3 gallons of lukewarm water per acre.

"You have to spray it so it's like rain, not mist," explained Pessereau, adding that the preparations are not used as fertilizer but "for the energy stored in the preparation."

The core of biodynamic farming, according to Pessereau, includes using no chemicals, using nine different preparations and using compost (six of the nine preparations are used to make compost).

Montes is the first to point out the difficulties of biodynamic farming, which include costs and timing challenges. "It's difficult to work with the calendar. That's the hardest part," said Montes, recalling a day when the calendar indicated a favorable time to apply horn silica preparation, which is made from crystals. The weather, Montes said, was "too hot so we backed off. We had to come back in cooler weather to avoid sun burning the vine."

Biodynamic farming has not been a simple undertaking, but Steiner's philosophy remains inspiring to Pessereau and Montes. "What Steiner was saying," Montes explained, "is to farm with conscience. We have to take care of humans and animals. We have to preserve the soil for future generations." ■

Chapter Seven *Organic Farming*

▲ "Connie" Corbett is transitioning his St. Helena winegrapes to organic. Corbett's and another family reside on the vineyard property.

▶ This northern Napa County vineyard is grown organically and positioned far enough from the creek to protect water quality and create a wide riparian corridor for wildlife.

Naturally complex issues

Organic farmers are typically motivated by strong personal convictions which are reinforced by a general public perception that organic farming is the most "natural" and therefore the most responsible way to farm.

Some growers, however, challenge this popular perception—they argue that an organic practice requiring more fossil fuel or more strenuous physical labor than a non-organic practice may not be sustainable. This argument points to the importance of looking at all the factors that may or may not sustain business, environment and worker well-being in a farming operation.

John Williams, whose Frog's Leap Winery and Vineyards in Rutherford were the first in Napa Valley to gain organic certification, believes that farming reflects deeply personal values. "Farming has always been and will always be an individual pursuit," said Williams. "I'm not a farmer myself, I'm a businessman. But the health of employees, the value of investments, the true cost of doing business—all of these grow out of how you treat your natural resources, be it people, land or water."

Organic certification, in Williams' view, is only a step toward the ultimate goal of a truly sustainable business— one that calculates for so-called hidden costs.[2] "I'm completely amused," Williams said, "by questions about the

▲ This small organic vineyard development preserves the surrounding natural woodlands. Vineyard work is done by hand, rather than with farm equipment, so there is little disruption to the site.

cost of organic farming." Williams points out that conventional business provides no place on a balance sheet for a long list of costs that include a worker's nosebleed caused by pesticides, squandered water supplies or compromised water or air quality. Even without calculating for hidden costs, Williams says, "organic farming is still cheaper. That's the irony."

But the hidden costs of conventional farming do need to be accounted for. Williams, now in his 22nd year of business, is adamant that "unless you're willing to consider the long-term costs, you're not going to be in business very long."

CHAPTER NOTES

[1] Dave Whitmer, "Historical Evolution of Agriculture in the United States", presented at *"Reunion Technique,"* Domaine Chandon, June 17, 1998.

[2] The concept of hidden costs in business was discussed by Paul Hawken, in his book *Eco-Commerce*.

PHOTO CREDITS

Juliane Poirier Locke: pages 118–119, 120, 121, 122, 124, 125, 127, 128, 129, 130, 131

Chapter Eight

▲ These local children[1] were paid farm workers in St. Helena orchards, after Prohibition forced some growers to take out vineyards and plant prune trees on Spring Mountain. Photo circa 1927.

People
Personal and social responsibility

At a Napa Valley winegrowers meeting held in August 1883, everyone reportedly agreed when a member—prompted perhaps by exasperation—announced, "Winegrowers[2] should live a thousand years to learn all about the business."[3]

About 120 years later, another grower said, "It can't be learned in a lifetime. Things keep changing and you have to be learning all the time."

While some sentiments about farming may not have changed since the 19th century, the political scope of the work seems to be expanding. Growers have little choice but to be aware that farming is no longer strictly a farmer's business.

Chapter Eight: People

Chapter Eight *People*

▲ This Yountville vineyard is bordered closely by homes. The vineyard manager posts a phone number so neighbors can call in their concerns.

Winegrowing has an impact upon countless ecological processes, from the spawning of migratory fish to the varieties and number of microscopic soil inhabitants. Vineyard practices are linked to laws made in Washington, D.C., in Sacramento and in Napa, and they are linked to the thoughts and feelings of neighbors—whether they wake up to the sound of a tractor at 3 o'clock in the morning or travel from their homes in Mexico to help pick the grapes.

Talking to the neighbors

As homes get closer to vineyards, and a curious—sometimes fearful or irritated—public looks over the fence to see what's going on, growers are expected to explain themselves and their practices. Growers who haven't bothered to do this in the past may be soon pressed to polish up their social skills.

Wine industry organizations are encouraging growers to talk truthfully to neighbors and inform them of everything, from what is being sprayed to when they can expect to hear tractors or wind machines during sleep hours. It is a routine practice at

▶ Hugh Davies (at far right) talks to a watershed tour group at Schramsberg Vineyards.

134 *Vineyards in the Watershed*

the vineyard and winery of Spottswoode in St. Helena, for example, to inform neighbors ahead of certain procedures *(see page 146)*.

Talking to neighbors is not just courteous but pragmatic. Due in part to public complaints made to local agricultural commissioners (who in turn inform the state Department of Pesticide Regulations), ground-level communication is on the agenda of several grower organizations along with the goal of improving pesticide applications to avoid over-spray and drift incidents.

While the California Association of Winegrape Growers (CAWG) is pushing to improve herbicide and sulfur application practices[4], it is simultaneously distributing a manual on how to communicate with neighbors.

"Most people just want to know the facts, so make sure you're providing them with the truth," advises the CAWG manual. "Tell it like it is, not like you want it to be."[5] A practical communication guide tailored to Napa County is now in the works for local growers.[6]

Seasonal workers—the other neighbors

Talk, on the other hand, can sometimes be cheap when it comes to labor relations. Next to environmental issues, what draws the most public interest to the farming industry is the seasonal farm worker.[7]

How farm workers are treated—including what they are paid and where they live while they are here—is a complex and politically charged issue. It is also central to any discussion of agricultural sustainability.

The living conditions for farm workers in Napa County have been well documented—sometimes dramatized—by the media. Between 3,000 and

◀ Napa County Agricultural Commissioner Dave Whitmer routinely encourages neighbors and growers to talk to one another.

◀ Vineyard manager Alfredo Gonzalez trains vineyard workers in the proper installation of a silt fence.

Chapter Eight: People

Chapter Eight *People*

▲ Workclothes dry in the sun at a farmworker camp.

4,000 workers come for the harvest in Napa County, where during the harvest of 2000 there were an estimated 286 (legally licensed) beds available in farm worker camps.

There used to be fewer grapes to pick and more beds available to workers. But as more vineyards have gone in, more camps have been closed or transformed into something else entirely. Where two labor camps once stood in Rutherford, for example, there are now luxury homes. The former site of a farm worker camp in Oakville is now a gym.

The loss of farm worker housing

Housing for farm labor used to be the responsibility of the grower, who hired workers directly and was more or less responsible for their board and keep while the work was being done. Then vineyards started expanding and labor camps started disappearing.

In 1975, there were 15,000 acres in grapes and 29 privately owned and operated camps; in 2001 there were over 37,000 acres in grapes and seven private camps. Camps have closed because the structures were condemned, because the buildings were needed for winery offices, because the camp was in the way of winery expansion and because neighbors complained about the noise and bad habits that emerge when a lot of men have to live together for seven months in close quarters.[8]

Labor camps also closed because they conflicted with the growing tourist trade, which was not the only business pressure to remove labor camps. The demise of private camps was in large part due to the industry shift toward corporate practices—housing was not profitable and therefore not pursued as a business practice.

In some cases where worker pay was increased, families were moved here or workers became citizens, on-site housing became less important at the time.

In situations where there were more workers than beds, employers had to decide which workers would be housed. The feelings and perceptions of those who did not get subsidized housing then became a source of tension that could affect worker relations.

Other difficulties that come with labor camps include the proper housing of women workers or, typically in the absence of women and families, the problems of social isolation.

Sustaining a commitment to labor

The few who continued to take care of workers demonstrate that private worker camps can be a sustainable part of winegrowing. The Wood family has been operating a farm labor camp near Rutherford since 1946.

The Solari family has kept their worker camp open since the late 1950s.

Reuben Oropeza, the county's inspector for the worker camps, believes that even though owners subsidize private camps, the payoffs are worth it. Oropeza, who grew up in a labor camp, says the benefits of housing workers always outweigh the costs.

"Believe me, it's a money-loser for (owners)," said Oropeza about private labor camps. "But they get rewarded in good employees."

Robert Mondavi was among the growers who did not close his camp. In terms of good treatment of workers, "Mondavi was the leader of the pack," said Oropeza, adding that the county now runs the Mondavi camp because most Mondavi workers are able to afford conventional housing and because the company work force is mostly permanent workers.

Andy Hoxsey, Oropeza said, is a grower who gives his workers a good deal. At the camp at Yount Mill Vineyards, harvest workers pay only $6.50 per day for meals and lodging. "He (Hoxsey) is always going to have good workers because he treats them well," said Oropeza. "He's one of those growers who takes care of his workers."

Most year-round workers at York Creek Vineyards have been able to purchase their own homes, and the private camp is kept open as a convenience for the few seasonal workers who still live in Mexico. Beringer Wine Estates, California Grapevine Nursery and Niebaum-Coppola Winery also operate other private camps.

Currently, the privately operated camps in Napa County house a combined total of 118 people, while the three publicly operated camps house a total of 176.

For about $10 a day workers get three (Mexican style) meals a day, along with a bed and showers at the

▲ Community volunteers help build the foundations for the yurts, temporary farm worker housing installed near Yountville.

Chapter Eight: People

Chapter Eight *People*

public camps. They also get direct access to social services ranging from English tutoring to AIDS lectures. The Highway Patrol routinely explains local traffic and alcohol laws to camp residents, and, most recently, workshops at the camps are training farm workers to identify the glassy-winged sharpshooter.

On their own

During his 18 years in Napa County, Oropeza has observed that when some laborers sleep in cars, under bridges, on the porch of the St. Helena Catholic Church and other places, it is typically because either the camps are full or because the workers don't want to follow the camp rules, particularly the no-alcohol rule. "There will always be small crowds that want to be on their own," Oropeza said.

Others who choose to go their own way include some homeowners who ignore the law and make money by renting sleeping space to workers in garages and basements. At one home in St. Helena, the county documented as many as 60 people illegally renting bed space. This practice is carried out at residences in every city in the county.

Meanwhile workers live where they can. It is obvious that more housing is needed, but while there is widespread community support for more permanent farm worker housing, the response of homeowners is to some degree predictable—those who may support worker housing in theory do not wish to have it in their neighborhoods.

But even if land were readily available, there is some question about the environmental sustainability of creating permanent housing for thousands of workers who already have permanents homes in Mexico and who are here only certain months of the year. Models such as the "tent city" at Yountville may be more economical and make a smaller ecological footprint[9] than permanent buildings.[10]

The growth of a year-round work force

The farm worker housing problem gets most public attention during the harvest season, when people are observed practicing "urban camping." However, the farm worker housing problem is an extension of the countywide shortage of low-income housing. (Blue-collar and white-collar workers alike can find little affordable housing in a county where land can sell for $100,000 an acre.) While it gets public attention mostly during harvest season, the shortage of adequate farm worker housing is a year-round problem.

Farm workers are typically either year-round or seven-month employees. For the past 10 years, the local farming industry has shifted toward a year-round work force. Beginning around 1989 phylloxera resulted in an estimated 75 percent of vineyards needing to be gradually replanted. These large-scale replanting operations shifted the peak work force needs to spring rather than fall. Concurrent with this era, the switch to high-density plantings and trellising also called for more labor in the spring. As replanted vineyards become stable the labor needs may likely shift again.

Because of the seasonal nature of farming, workers can expect routine layoffs throughout the year. In some cases, workers have health benefits. There is also a range of jobs, with workers getting paid more for specialized skills.

Worker wages

Vineyard workers can make more money for their labors in Napa County than in other grape growing counties in California. But even while the prices commanded by Napa County winegrapes have in some cases tripled in less than two decades, the wages paid to farm workers have improved in some cases and gone down in other cases. In some cases, labor contracting has not been good for farm worker wages *(see case study, page 140).*

▲ A harvest worker in a St. Helena vineyard.

Cabernet Sauvignon grapes in the 1987 harvest yielded 16,000 tons at $1,096 per ton; in the 2000 harvest, 42,925 tons sold for $3,168 per ton.[11] A Mexican farm worker during the 1980s could expect to work from March through October and return home from the Napa Valley with as much as $8,000. A laborer doing the same work in 2001 took home about $7,000. Some workers interviewed for this book said they saved about $4,000 in 2001, and a few (who changed employers several times throughout the year and thus did not work full time) said they took home only between $1,500 and $2,000. Income depends on the ethics and resources of the employer, the amount of work to be done and the aptitude of the worker.

Chapter Eight: People

Case Study: *Francisco Ramirez, Mexican farm worker*

▲ Francisco Ramirez

Francisco Ramirez is a seasonal farm worker who finds Napa County a better deal than working in Fresno. Although some Mexican workers are exploited under the labor contracting system, Ramirez works for a contractor who gives him steady work at a decent wage.

▲ Angel Calderon, manager for the county-operated farm worker camp in Calistoga.

Francisco Ramirez grossed $10,000 working in Napa vineyards between March and October of 2001. After taxes and living expenses, he was able to send about $6,500 to his family in Oaxaca, Mexico.

Harvest is almost over when Ramirez, sitting in the dining hall at the farm worker camp in Calistoga, turns away from a late dinner to explain—through a translator—that it is "difficult to find a place to live" in Napa, but the wages are better than what he got in Fresno.

(Although no formal study has been conducted, Napa County farm wages are thought to be the highest of any agricultural operation in the country.)

Ramirez, 59, has been harvesting grapes in California since 1978. Five years ago in Fresno he made $4.25 an hour and paid about $80 per month to share a one-bedroom apartment with nine other people. Food expenses were extra.

Traveling and working in a family team with his brother and two sons, Ramirez now works in Napa County for Mario Bazan, who pays Ramirez $8.75 an hour and gives him steady work seven months of the year.

Someone who knows Ramirez describes him as a reliable worker with a good attitude. He also happens to be fortunate in his employer. As the labor contracting industry in Napa County expands, there is more opportunity for workers to be exploited—a situation that challenges the social sustainability of winegrowing.

Angel Calderon, manager of the county-operated farm worker camp in Calistoga, hears from the workers about what is going on in the fields. Calderon says some contractors pay fairly and treat workers well, and some do not. A contractor may charge a vineyard company an hourly rate for labor, paying the workers significantly less per hour. Some contractors charge workers for transportation to the job site. Calderon told of a worker who, after getting a day's wages from a labor contractor, paying his $10 rent at the farm workers camp and the $6 transportation fee charged by the contractor, had only $15 to show for 10 hours of work.

A large Napa Valley vineyard company that paid between $110 and $120 per ton to workers in the 1987 harvest now uses a labor contracting service that paid workers $75 per ton in the 2001 harvest. The contractor also charged each worker $12 for the picking pan used to collect the grapes.

Under the labor contracting service, workers have greater opportunity to exploit fellow workers. Migrant workers most often work in teams called *cuadrillas*. Contractors will pick a team by leader—usually the one who speaks English—and the leader gets the paycheck. The team leaders can then make money by skimming something for themselves before paying fellow *cuadrilla* members.

Ramirez is happy with his contractor-employer, but some workers would prefer having no contractor system. One of them is Ramón Cardenas, a 46-year-old worker from Jacona, Mexico, who has worked in Napa for 18 years. Ramón changed jobs several times in 2001 and took home only about $4,000. He worked for a low-paying contractor and a well-paying contractor, but said the best pay is always when he works directly for the owner-grower. ■

▲ Ramón Cardenas, seasonal farm worker from Jacona, Mexico.

▲ Francisco Ramirez (right) with his son, Oliveras, both seasonal farm workers in Napa County.

Chapter Eight: People

Chapter Eight *People*

Currently in Napa County, a seasonal farm worker can make between $8 and $20 an hour for harvesting grapes, with an average of about $13 per hour.[11] Pruners and grafters can make as much as $15 per hour. Some companies offer premiums for night shifts, profit sharing, bonuses and cash incentives for performance. The maximum hourly rate for a harvest worker is $22 per hour, but the hours are irregular. A worker may work three to six hours a day for a few days, then go without work for several days, waiting until the next vineyard is ready to be picked.

Nord Coast Vineyard Service is one company that adjusts pay to picking conditions. Under good conditions where fruit is plentiful, workers in 2001 got paid $75 per ton; in blocks where the fruit had been thinned and was harder to pick, the pay rose to $100 per ton or $12 per hour— whichever was greater.

As in any other business, vineyard employers have to balance costs in determining the wages paid to employees. Workers can and do leave agriculture for higher paying trades such as construction. To be sustainable, a vineyard needs workers who are skilled, dependable and satisfied with the conditions and pay of their work.

Taking responsibility

Although the labor situation is a highly complex one, there is a great deal a grower can do to influence the quality of working and living conditions for workers. Growers using a labor contracting service can and should impose restrictions upon that contractor's management or mismanagement of employees. It is not a sustainable practice to look the other way when social equity is being compromised.

Down by the riverside

In addition to worker issues, criticism of agriculture focuses largely on farming practices that affect the plight of the environment—another topic dear to the public heart.

Agricultural practices are often improved as the result of farmer education, land ethics and personal initiative. In other instances, farming changes due to public opinion, frequently influenced by public interest groups using the media to raise public consciousness about critical environmental and health issues. The world ban on methyl bromide, for instance, was accomplished via public awareness campaigns backed by scientific studies.[13]

Conflict, however, between environmentalists and the agriculture industry is legendary on local, regional and national levels. Litigation—justifiable or otherwise—is often part of those battles, at a cost of time, resources and trust among stakeholders.

But members of the farming industry, public interest groups and environmentalists, in some circumstances, have shared a paradigm shift—from conflict to cooperation.

Rather than condemning all farmers for what the minority is doing wrong, some environmentalists credit worthy farmers for what is being done right. In Sonoma County, for example, river advocates collaborated with winegrowers on a "fish friendly farming" program intended to enhance sustainability in vineyards and to help promote a healthy watershed.

Local partnerships

Other examples of cooperative relations among stakeholders can be found in the efforts of national and international public interest and environmental groups[14] and a number of major public interest funding organizations[15] that support sustainable farming.

In Napa County, the Resource Conservation District (RCD) and the Napa Sustainable Winegrowing Group (NSWG) are leading a collaborative effort—with local vintners, farmers, environmentalists and others—to create a "green" certification program for vineyards. These stakeholders are working to devise a program—founded on baseline watershed studies now in progress—that will assist farmers in raising their standards of stewardship.

In essence, the program will provide a means by which growers can evaluate their farming practices in context with the general status of the watershed. Further, the program will provide incentives for doing the right thing, apart from the pressures exerted (or not exerted) by government regulation.

Some growers who practice good land management because of a personal land ethic have contributed to the community by their involvement in land stewardship groups. Even those who do not espouse a personal land ethic but nevertheless participate in a stewardship group are doing the right thing, regardless of motive.

▲ Jon Kanagy, Juan Cardenas and Phill Blake discuss restoration strategies in Dry Creek, early fall of 2001.

Chapter Eight *People*

A key to stewardship progress is an individual's ability to recognize that he or she is part of the problem.

Stewardship groups

Landowners have been part of the local stewardship efforts that have resulted in restoring tributaries to the Napa River. Stewardship groups have been organized on Huichica Creek, Carneros Creek, Dry Creek, Sulphur Creek, Salvador Channel, Redwood Creek, Garnett Creek and Hopper Creek.

Leigh Sharp, who coordinates stewardship groups on behalf of the RCD, describes the stewardship process as an alternative to changes brought about by regulation. "People don't like to be told what to do," said Sharp, explaining that a stewardship group organizes itself around the interests and wishes of the members. "Rather than a top-down," Sharp said, "it's really a bottom-up process."

Stewardship groups that are started from outside the sub-watershed usually meet with less than success. To be sustainable, groups need committed members. "Generally stewardships are most successful when they are started by landowners," said Sharp "You can't make somebody be interested in this."

Belonging to a stewardship group is "a way of getting yourself out of yourself and into the community," said Lee Hudson of Hudson Vineyards in Carneros. "The Huichica Creek process was a positive philosophical adjustment for me—to look at my property as part of the watershed." Hudson now belongs also to the Carneros Creek stewardship group and is pursuing an interest in stream flow monitoring.

Some landowners have found stewardship groups helpful in determining creek water flows, improving fish runs, replacing vegetation and restoring creek banks—preventing creek bank failure and from claiming valuable vineyard land.

One agency official with extensive watershed stewardship experience observed that a key to stewardship progress is an individual's ability to recognize that he or she is part of the problem and therefore has some responsibility for working—cooperatively with others—on the solution.

◀ Opposite page: Straw bales, jute netting and irrigation line are basic materials used in creek restoration.

Partnerships—new ways of relating with agencies

Ann Thrupp of the Environmental Protection Agency (EPA) observes that among the California agriculture groups with which she works, winegrowers are most willing to share information. Although there is the usual industry bickering that goes on, she explained, there are efforts to improve relationships among those who work in agriculture, government, science and academia.

The EPA itself has shifted toward a more cooperative mode of operation, evidenced by the agency's Agricultural Initiative program, now in its fourth year. Thrupp, who directs the program, says the goal of her unit is "to support sustainable agriculture in the transition to reduce reliance on pesticides."

Toward that goal, the EPA offers partnerships in the form of cost-sharing or grants as incentive for landowners to do the right thing. Partnerships "help people develop proactive ways of dealing with environmental issues before they get hit with enforcement," said Thrupp, who warns against using partnerships as the latest form of green washing.

"People have to practice what they preach," Thrupp said. "If they sit with the 'green people,' it doesn't mean they're not going to get hit if they pollute. Some partnerships are just chitchatting, but the ones that actually work," she explained, "include the Lodi Woodbridge Winegrape Commission, the Sonoma County Grapegrowers Association and the Napa Sustainable Winegrowing Group."[16]

◀ The Napa River Ecological Reserve near Yountville is the only significant acreage on the Valley floor that remains undeveloped.

Chapter Eight: People

Case Study: Spottswoode Vineyard and Winery
Napa River/Sulfur Creek/Spring Creek watershed

▲ Beth Milliken

The shrinking boundaries between agriculture and suburbia have placed growers in the role of communicator— a job that includes notifying and educating the public about vineyard practices.

"We're surrounded," said Beth Milliken of Spottswoode Vineyard and Winery. Outside on the porch, a group was gathered for a scheduled tour of the organically grown vineyard and the historic winery building in St. Helena. But Milliken wasn't talking about the visitors—she was talking about the homes in the neighborhood.

Inside the city limits, on property surrounded on all sides by houses, the vineyard and winery of Spottswoode must operate under the eyes of residents on Spring Street and Hudson, Sylvaner, Reisling and Madrona avenues.

Being so close to her neighbors, Milliken has made a point of communicating with them as much as possible. For almost a decade she has routinely kept neighbors informed about farming operations, such as when they might expect to hear wind machines or when tractors will be out applying sulfur.

"We write a few letters a year to neighbors," said Milliken. "We're not asking permission—more just informing and educating them as to our farming practices and our beliefs." The mailing list is up to 69 recipients and includes the president of The Mennon Environmental Foundation, which funds local environmental projects.

"In the old days there was no communication between growers and neighbors," Milliken said. "I think it's people's right to know and it's our obligation to keep neighbors informed." She added that the burden of communication falls on the grower. "The more information you can give people the better."

Spottswoode issues a newsletter to neighbors once a year and keeps an open invitation for neighbors to call with questions. So far, there are no conflicts with any of the neighbors. In contrast to some small watershed areas, neighbors have also been coop-

erative about a creek restoration project now in the planning stage.

There has been bank erosion on Spring Creek, which defines the southern boundary of the vineyard and empties into Sulphur Creek. Efforts that individual property owners made to prevent erosion had, in some places, contributed to increased erosion elsewhere. Milliken hired a creek restoration consultant and invited neighbors to discuss approaching Spring Creek "as a whole and not piecemeal."

The project is waiting on funding.

"We have relatively good relations with our neighbors, especially considering how many we have," said Milliken. "Most vineyards don't contend with as many neighbors as we do." She appreciates the cooperative nature of her neighbors, and many of them don't mind living so close to a vineyard and winery. "They appreciate the view and the organic farming we do," said Milliken. "It helps keep their property values high." ■

▲ Milliken and her neighbors are working together on a restoration plan for Spring Creek. Sites such as this one will benefit from revegetation and other "soft" techniques to replace ineffective "hard armor" approaches used in the past.

◄ Spottswoode Vineyard and Winery, at left, is part of several densely populated neighborhoods in St. Helena.

Chapter Eight: People

Chapter Eight *People*

Landowners and vineyard managers who would confuse the right to farm with the right to ignore human needs or wildlife requirements are pursuing a costly nostalgia.

What people want

Partnerships have to work against a tradition of conflict and within the limitations of basic human nature. As former opponents agree to sit down and talk, even work together to protect the watershed, an informed public is watching what the farmers are doing and asking why.

While wine marketers may salivate at the chance to "sell" sustainable farming, the test of whether an operation is actually moving toward sustainability is in part determined by the quality of relationships between growers and their communities and between farmers and the people who work for them.

In a county as small as Napa it is particularly true that landowners and vineyard managers who would confuse the right to farm with the right to ignore human needs or wildlife requirements are pursuing a costly nostalgia. Very little separation remains between agricultural and urban pursuits. Since the population keeps growing and more than a few vineyards are surrounded by homes, the curiosity about what farmers are doing and why is likely to grow also.

Farmers have always needed to know a lot of practical science. But if agriculture is to be sustained through the next century, farmers will also need to know—and talk to—their neighbors.

Meanwhile those neighbors and people in the rest of the state and the country will keep watching to see how sustainable farmers tend to the watershed and to the living, earning and working conditions of the farm worker.

▶ Ignacio Cardenas, Arturo Guerrero and Javier Sandoval—workers for Nord Coast Vineyard Service—prepare room for a willow revetment in Dry Creek, a tributary of the Napa River.

Case Study: Shrimp and stewardship—
A group effort on Huichica Creek

Only a scattering of Pinot Noir grapes still hang from the vines as Vince Bonotto inspects a sediment trap in a Carneros vineyard. Where a vertical pipe sticks out of the ground beside a dike fashioned from straw bales, Bonotto kneels and points to what looks like dried mud in the bottom of the pipe.

"Time to clean this out," he says, peering at the accumulated sediment. The device catches rainwater runoff from the vineyard, filters out the sediment then releases the water—at a slower travel rate—into the stream downhill of the vineyard.

A decade ago, when Bonotto planted Devaux in Carneros—a 112-acre vineyard of Mumm Napa—the watershed management plan devised for his site called for the planting of native grasses as cover crops and very broad setbacks from the creek. In addition, Bonotto spent $60,000 in erosion control devices that are routinely maintained as watershed conditions are routinely monitored. He is responsible for protecting a tiny organism that lives in the waterway.

Huichica Creek is one of the last creeks in the Northern Hemisphere that supports the California freshwater shrimp *(Syncaris pacifica)*, an endangered species. The shrimp like to spend summers right in the part of the creek that flows through the site of the Devaux vineyard.

The setbacks and the native cover crops are the most significant water quality protections at the vineyard site. The setbacks keep farming operations from disrupting the riparian area, and the native cover crops control erosion while improving the soil's ability to absorb water.

"The management has been good for the creek," said a local government representative, adding that Huichica Creek is "in better condition than it was." Prior to the Devaux vineyard, the property had been used for livestock grazing, which had contributed to erosion, water quality degradation and the decline of riparian vegetation.

▲ Vince Bonotto manages the Devaux vineyard, along Huichica Creek in Carneros, where he has learned a lot about protecting shrimp habitat.

▼ The vineyard was planted without disturbing this native oak or the ephemeral stream beside it.

Chapter Eight: People

No horses or trucks cross through the creek anymore. And the seedlings that sprout up in the shade along the banks of the creek are occasionally eaten by deer, but not devoured wholesale by grazing livestock.

In figuring out how to protect the watershed and the shrimp's habitat while still getting their farming done, stakeholders in the Huichica Creek watershed 10 years ago did something original—they cooperated.

Cooperation is still a new concept in land-use negotiations. When it comes to regulating public resources, conflict is not unusual between landowners (who must abide, even if the rules aren't practical) and regulatory agencies (which usually make the specific rules, based on laws), and between one agency and another (each of which operates according to its own codes and agendas). But when the Huichica Creek stewardship group got under way, participants devised their own, voluntary management plan. Regulatory agencies approved the plan.

It was not always smooth going. "Some of us said, 'Yeah, let's do this,' and some of us had to be dragged kicking and screaming," Bonotto said. "There was some skepticism about whether regulatory agencies would listen, but they did. It was a learning process for everyone."

Bonotto knows a lot about the California freshwater shrimp, although —ironically—he has never seen one.

▶ An abandoned winery, in ruins at the Devaux property, was built about 20 feet from the creek. Because Huichica Creek changed its path over the years, it now flows right alongside the old winery door.

◀ A straw bale dike slows and filters runoff before it reaches this inlet pipe, which traps sediment.

hydrology combinations, and (5) completing a water survey to establish the average runoff from each watershed section to help landowners and managers stabilize stream flows.[19]

He also knows the portion of creek that runs along his vineyard is "textbook habitat" for the endangered organism. "We didn't do any tree removal near the creek," Bonotto said. "Shade is important to the shrimp, and root balls are where they spawn. The juveniles cling to the roots for protection."

The Huichica Creek stewardship group inspired a countywide watershed management guide[16] and became a model for other watershed collaborations and trainings around the country.[18] The Huichica Creek stewardship group accomplishments were recognized in a national watershed publication, listed as: (1) enlisting 63 of the 70 local landowners to participate, (2) restoring and stabilizing 800 feet of stream banks, (3) planting at least 10,000 trees, (4) planting four demonstration sites to show the suitability of different cover crops to various soil-hydrology combinations, and (5) completing a water survey to establish the average runoff from each watershed section to help landowners and managers stabilize stream flows.[19]

But measuring the success of subsequent stewardship groups against the success of the Huichica Creek group may not be entirely fair, since the problems on other tributaries are often more numerous and complex.

Bonotto explained that the Huichica Creek group was "unique because it was mostly growers farming on or living near the creek." The group addressed water quality and quantity as well as habitat health for the shrimp. Bonotto now belongs to the Carneros Creek stewardship *(see page 46)* where there are "more problems" and the constituency includes the Resource Conservation District (RCD), the Natural Resources Conservation Service (NRCS), commercial developers, residents, growers and others.

"We have groundwater issues and erosion issues," said Bonotto. The group also has steelhead issues. "I know we'll be able to assess the watershed and do wildlife and habitat studies and develop plans for those aspects, but whether we can solve the groundwater problems, we don't know yet." ■

Chapter Eight *People*

▲ Top: This revetment is constructed entirely with willow posts, branches and jute twine. The structure is irrigated until the willow sprouts and takes root. At the base of the structure, small willow cuttings help stop stream flows from cutting under the revetment.

▲ Willow posts, strategically placed in this creekbed, will take root and become trees to provide channel stability, shade and habitat for riparian species.

CHAPTER NOTES

[1] From left to right: Jack Wilson, Dick Myrick, Mary Montelli, Phillip Conradi, Jule Myrick and Gertrude Conradi.

[2] Many believe the term winegrower is of recent coinage, but in fact dates at least to the late 1800s.

[3] Comment by H. A. Pellet, reported in *St. Helena Star* Supplement, August 14, 1883. The comment followed an inconclusive discussion in which growers pondered the origins and prevention of red leaf disease.

[4] The Pesticide Management Alliance (PMA) is a network of state winegrowers, spearheaded by the California Association of Winegrape Growers (CAWG) with industry funding and a grant from the Department of Pesticide Regulations (DPR). The goal of the PMA is to educate both growers and the public about improved farming practices.

[5] Anne Chadwick, *The Winegrape Guidebook for Establishing Good Neighbor and Community Relations*. Sacramento: California Association of Winegrape Growers, 2000. p. 3.

[6] The Napa Valley Vintners Association will publish a communication guide in 2002.

[7] Whereas migrant farm workers traditionally follow crops, the seasonal workers in Napa County typically have homes in Mexico and return to Napa to work for either a seven-month period each year, or just for the three-month harvest period.

[8] Juliane Poirier Locke, "Farmworker Housing Update Part I: Where have all the labor camps gone?" *St. Helena Star,* Nov. 9, 2000: p. 1.

[9] Ecological footprint is a term coined by Mathis Wackernagel used to describe the combined natural resources consumed in a given activity.

[10] The yurts and "kitchen" tent used at the Yountville site, because they do not require asphalt or cement foundations, allow the ground to remain permeable and thus do not significantly increase storm water runoff rates. The tents are removed at the end of each harvest season, and there is no year-round maintenance required.

▲ Roses and a statue of the Virgin de Guadalupe adorn the walkway at the farm worker camp in Calistoga.

[11] The Napa County Department of Agriculture 1987 and 2000 Crop Reports, respectively.

[12] Figures taken from a private, independent wage survey conducted in 2001.

[13] Agricultural lobbyists were able to extend, for several years, the deadline for banning methyl bromide, and an equally toxic—though not ozone-depleting—replacement product, methyl iodide, is awaiting EPA approval. Methyl iodide, if approved, will not likely be used in sustainable farming operations.

[14] These groups include but are not limited to Pesticide Action Network North America (PANNA), which is an umbrella organization for groups including Californians for Alternatives to Toxins (CATS) and Californians for Pesticide Reform (CPR); the Natural Resources Defense Council (NRDC); the World Wildlife Fund (WWF); and the Nature Conservancy.

[15] In California, the Funders Agriculture Working Group includes over a dozen funding organizations, public and private, which promote and fund sustainable agriculture.

[16] The Napa Sustainable Winegrowing Group is a partnership among UC Davis researchers, the UC Sustainable Agriculture Research and Education Program (UC SAREP), the Napa County Department of Agriculture, the Napa County Resource Conservation District, the USDA Natural Resources Conservation Service (NRCS) and others.

[17] *The Napa River Watershed Owner's Manual: An Integrated Resource Management Plan.* Napa County, 1994.

[18] Juliane Poirier Locke, "State of the Watershed (Part VII): River advocates laud cooperative efforts," *St. Helena Star,* May 11, 2000: p. 1.

[19] "Agriculture and The Environment: Information on and Characteristics of Selected Watershed Projects." U.S. General Accounting Office (GAO) Report to the Committee on Agriculture, Nutrition and Forestry, U.S. Senate. June 1995: p. 23.

PHOTO CREDITS
Collection of Tom Wilson: pages 132–133
Juliane Poirier Locke: pages 134, 135 *top*, 136, 140, 141, 143, 144, 145, 146, 147, 148, 149, 150, 151, 152
Astrid Bock-Foster: page 135 *bottom*
Richard Camera: page 137
Collection of de Leuze Family Vineyards: page 138, 139, 153 *bottom*
Angel Calderon: pages 137, 153 *top*

▲ No complaints about farming from this house. It's the vineyard manager's residence, adjacent to an organically grown vineyard.

Chapter Eight: People

Appendix I: *Site Assessment*

Land and Resources Inventory

Successful sustainable farming begins with a plan—and several good maps. Putting sustainable principles into practice will be dictated to some degree by the physical features of the land. It is essential to know your terrain, soils, microclimate and property history in order to effectively manage a sustainable vineyard.

Begin with regional topographic maps, such as a **U.S. Geological Survey Quadrangle**, or a detailed **map of the watershed**. The USGS "quad" maps supply general detail on land slope, aspect, drainage patterns, surface water resources, surrounding land features and a bird's-eye view of your location in the watershed. The county tags certain watershed lands as sensitive, by virtue of vegetation, land instability, natural habitat type, runoff patterns or potential presence of federally listed endangered species. Federal, state and local government agencies generally map these resources on quads, or maps of a similar scale and format.

Soil maps are also important evaluation tools. Watershed-scale soil maps developed by the USDA Natural Resources Conservation Service are available in hard-copy form or digitally accessed at *www.ca.nrcs.usda/mlra.napass/napass.html.* The accompanying land descriptions and soil quality interpretations help explain the genesis of the landscape as it relates to proper uses or limiting features. Climate data and historical use are also found within the survey.

The county assessor's **parcel maps** and accompanying information provide general detail on parcel boundaries, easements and land acreage. A recent addition to the Napa County web site provides parcel line boundaries overlaid on an aerial photograph of the property and vicinity, at *http://co.napa.ca.us/internet/content/gisweb.* Go to "Mapping Applications," and pick "parcel-based maps."

Aerial photos are key to assessing land features including vegetative cover, road locations, general vineyard health, soil drainage, runoff patterns and sun angle/shade overlay. Historical aerials provide a snapshot-in-time look at soil quality, landscape stability, vegetation patterns, flood potential or drainage problems.

For vineyard planning, a large-scale **aerial map** of the vineyard and adjacent lands should include details on vineyard block layout, avenue and headland locations, block-by-block acreage features and vine row layout. Once the basic map is prepared, overlay information might include variety/clone types, rootstock selections, cover crop species and vine density.

An **underground features map,** for devices such as irrigation pipelines, erosion control collection pipes and subsurface drainage tubes, will save time and expense when future excavations are needed.

Concepts for this appendix developed by the Napa Sustainable Winegrowing Group. Text by Phill Blake.

Logistic and Legal Parameters

County, state and federal laws sometimes place special parameters upon, and require surveys or studies for, the following features:

Water Features
- water rights (junior & senior)
- well water
- recycled water
- watershed runoff impoundment (reservoir)
- water quality
- water quantity

Biological Features
- wetlands, vernal pools
- sensitive, threatened and endangered plant and animal species
- pest history/potential pests
- riparian features

Cultural Features
- historical features
- archeological features

Appendix I: *Site Assessment*

An example site

Parcel/zoning map

Aerial photo

Historic photo

Zoning Features
- zoning
- adjacent properties
- rural/urban interface
- easements
- property borders

Biological & Physical Features
- vegetative cover
- roads & structures
- vineyard health
- drainage patterns
- neighboring land uses
- rural/urban interface

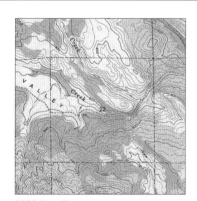
USGS "quad" map

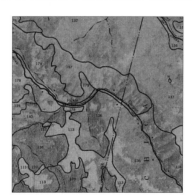
Soil map

Watershed map

Geological & Physical Features
- slope, aspect
- drainage, springs
- landslide history
- earthquake faults
- soil type(s)
- large & small drainage patterns

Appendix One: Site Assessment

Appendix II: *Farm Plan*

	January	February	March	April	May	June
Post-Harvest						
Disease Control						
Erosion Control						
Dormant						
Vine Assessment (pruning weights and visual assessment)						
Pre-pruning						
Finished Pruning						
Trellis Repair						
Weed Monitoring and Control						
Frost Protection						
Bud Break – Harvest						
		Disease & Pest Control				
		Frost Protection				
		Replanting/Training				
		Canopy Management (suckering, leaf removal, cluster thinning)				
			Irrigation			
Erosion Control						
Employer/Grower Training						

July	August	September	October	November	December
			Disease Control		
		Planting Cover Crops			
			Weed Control		
	Soil Amendment				
		Erosion Control			
		Fertilization			
			Experimental Review or Design		
				Vine Assessment	
					Pre-pruning
				Finished Pruning	
					Trellis Repair
					Weed Control
					Frost Protection
	Harvest				

Appendix Two: Farm Plan

Appendix III: *Erosion Control*

▲ Phill Blake developed the erosion control system on which Napa County's conservation regulations were based.

Planning for soil retention

By Phill Blake
District Conservationist, NRCS

Those who want to develop a new vineyard or replant an existing one should be aware that it's not as simple as ordering the vines and bringing in the bulldozer. For the past 10 years, Napa County has required growers to develop a detailed development or redevelopment plan before the first blade of grass is disturbed.

Wherever lands have a 5 percent or greater slope—that is, 5 feet or more of rise or fall in 100 feet measured horizontally—land cannot be cleared for planting or replanting unless the county first approves an Erosion Control Plan Application (ECPA) for the site.

What the county requires

The Napa County conservation regulations, commonly referred to as the "hillside ordinance," require practices that help control soil erosion, protect downstream lands and waters from siltation, and prevent clearing and development of sensitive riparian (streamside) habitats. The regulations also require development setbacks from streams on all lands proposed for new development, regardless of slope, a fact not known to all, because of the "hillside" moniker.

As a service to grower-applicants and the county, the Napa County Resource Conservation District (RCD) reviews all agricultural ECPAs. The RCD is not a county agency, but rather a special district established and overseen by local citizens to provide technical advice on a non-regulatory basis to landowners. The RCD also works with other agencies, such as the USDA Natural Resources Conservation Service (NRCS), to provide additional assistance and support for the local program.

How the process works

To begin the process, an applicant usually consults with the RCD, a private consultant or NRCS, then submits the ECPA to the county planning department. Planners check the application for completeness and for adequate environmental support documentation. The plan is then forwarded to the RCD for an on-the-ground consultation and fine-tuning with the applicant or applicant's consultant.

The RCD applies various technical modeling methods to ascertain the effectiveness of soil erosion control practices and stormwater runoff control measures. The RCD board of directors then notifies the county of

Erosion Control Plan Map

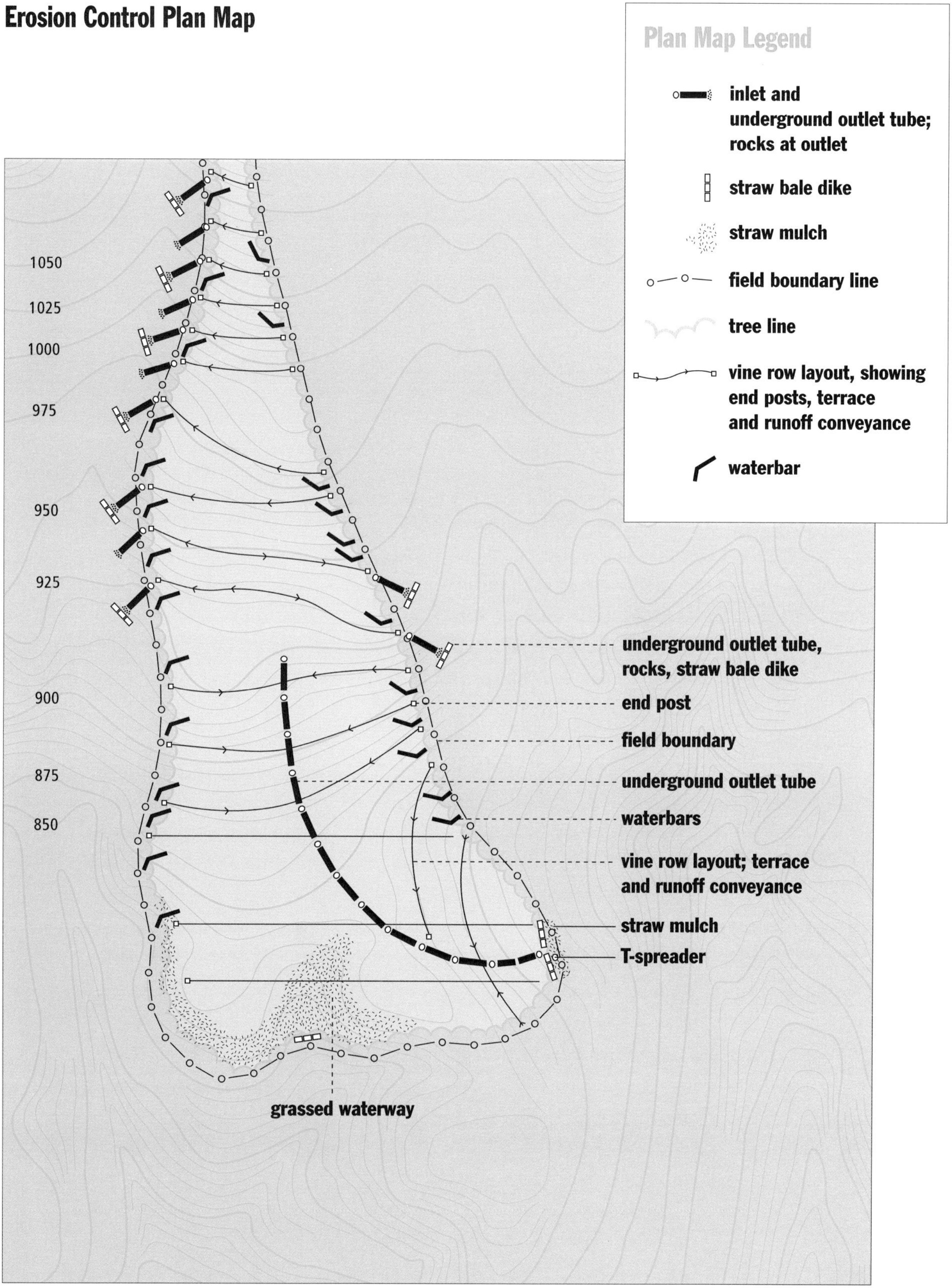

Plan Map Legend

- ○━━ inlet and underground outlet tube; rocks at outlet
- straw bale dike
- straw mulch
- ○─○─ field boundary line
- tree line
- vine row layout, showing end posts, terrace and runoff conveyance
- waterbar

Labels (right side):
- underground outlet tube, rocks, straw bale dike
- end post
- field boundary
- underground outlet tube
- waterbars
- vine row layout; terrace and runoff conveyance
- straw mulch
- T-spreader

grassed waterway

Elevations: 1050, 1025, 1000, 975, 950, 925, 900, 875, 850

Appendix Three: Erosion Control

Appendix III: *Erosion Control*

County regulations require practices that help control soil erosion, protect downstream lands and waters from siltation, and prevent clearing and development of sensitive riparian habitats.

▲ Straw bale dike

▲ Mid-slope diversion

their findings, which follows with final review of environmental documentation, prior to approval by staff or the planning commission.

The basic ECPA submittal package requires a detailed topographic site map, with soils information and specific delineation of vineyard development or replanting areas. Mapping must also note locations and details of erosion control practices, existing streams and required setbacks, and the layout of vine rows. Further details, such as an implementation schedule (from clearing to erosion control installations and planting of vines), are noted, as well as an accounting of locations, size and species of all trees to be removed.

How CEQA has changed the process

Since the county's original lawsuit settlement with the Sierra Club, planting a vineyard has become an even greater planning challenge for growers. Plans must now go through an evaluation process that uses the California Environmental Quality Act (CEQA) as the litmus test for approval. Depending on the size and scope of the project, as well as proximity to various environmentally sensitive watershed locales, required environmental documentation might range from fairly basic information submittals to full-blown detailed studies.

Because CEQA requires documentation of the potential "cumulative impacts" of the proposed action, ECPAs must now address both on-farm issues and the potential effects of the project on surrounding lands and downstream watershed areas.

For agricultural projects, typical environmental documentation includes such things as potential effects the vineyard may have on surface and

▲ Riser in sediment basin

▲ T-spreader

PHOTOS: PHILL BLAKE

groundwater supplies, sensitive wildlife habitats and wetlands, endangered species and natural stream flow and flooding processes.

Possible impacts of tighter regulations

Since the initiation of CEQA review processes, typical ECPA review periods have been extended from several weeks to several months or longer. There is hope, however, that processing time may be reduced, as the county revises the review process through recommendations of the Napa River Watershed Task Force. Future comprehensive environmental studies of the watershed may also be required to boost the county's confidence in issuing plan approvals under CEQA. Regardless of how we get to agreement on fixing the review process, vineyard expansion will likely remain slow, and the average size of most vineyard development units will be modest.

Restrictions, or perhaps outright prohibitions, on conversion of forests, oak woodlands and sensitive habitat areas is also a likely outcome that is already taking shape within the current plan processing structure.

Sustainable farming philosophy will certainly continue to challenge growers to look beyond the vineyard end posts and view farming operations decisions in their proper context as watershed management decisions. These increased restrictions on vineyard developments, together with the implementation of sustainable farming practices, will offer survival advantages to wildlife and reassurance to the public, who are increasingly demanding "green" returns on their investment in a bottle of wine.

Increased restrictions will offer survival advantages to wildlife and reassurance to the public, who are increasingly demanding "green" returns on their investment in a bottle of wine.

Appendix Three: Erosion Control

Appendix IV: *Soil*

The ground we stand on

Solid ground implies foundation and security, as a figure of speech. But in agriculture, ground that's too solid can mean disaster—the farmer's ground can't be foundational unless it's full of holes. Small ones.

A close look at good, fertile soil reveals a porous underground, where tunnels and pockets allow gas exchange, and through which water travels and roots reach into the busy lives of hundred of thousands of organisms. On the sites where soils have become compacted, there is often loss of those biological components that help turn mere rock dust into rich, life-promoting soil.

Soil controls how rain or irrigation gets distributed, stored, drained or infiltrated. It also regulates biological activities and molecular exchanges among solid, liquid and gaseous phases—all of which furthers the breakdown and conversion of old plant matter and the creation and growth of new plant matter.

Compaction

In Napa County, soil compaction in many places is preventing soil from doing its job. While good soil quality helps reduce erosion, sustain water supplies and maintain water and air quality, compacted soil cannot.

Compacted soil, when it's dry, can appear cloddy, or like concrete. After a rain, water often pools on the surface, evaporating rather than draining. In such condition, plants have difficulty getting nutrients and water, and they don't achieve good root penetration. Compaction can lower soil temperature, decrease the activity of soil micro-organisms and consequently decrease the release of nutrients needed by plants.

When storm water runs off compacted soil, it gains speed and causes erosion. The risk of compaction is greatest when soil is wet. Most often compaction occurs on vineyard roadways where vehicle weight compresses soil.

Infiltration

A well-managed soil with good structure and continuous pores will allow rainwater to move into the soil in a process known as infiltration. Healthy soil maximizes infiltration and allows rain to enter the ground continuously throughout a rainstorm.

When the soil takes in rain, it stores the water temporarily until it can be released slowly over time at the surface or carried underground to recharge ground water supplies.

Where infiltration is blocked or reduced by poorly managed soils, rainstorms more often produce ponding or runoff. The runoff can carry soil particles, fertilizers or pesticides into surface waters, causing problems for downstream neighbors, watershed integrity and animal life. Increased runoff contributes to flooding and to the land and property damages flooding can cause.

Keeping tillage to a minimum and keeping machinery off wet soils will help prevent compaction of the soil. To prevent soil crusting at the surface, increase soil porosity and help restore soils that have been compacted, cover crops are the single most effective tool. Where cover crops send their roots, the soil becomes more porous and more able to absorb and hold water. Cover crops also add organic matter—living and dead—to the soils.

Organic matter

Organic matter contributes to the infiltration capacity of the soil, preventing soil aggregates from breaking down under the impact of raindrops. If the soil particles were to break down, they would clog the pores through which rain travels in to the soil.

Formerly living plant and animal materials comprise the small organic portion of soils. When these dead things are well decomposed, they form humus, which is porous and accounts for the aroma particular to rich soils.

Soil requires the presence of organic matter in order to feed microbial populations, reduce erosion by keeping soil particles together, allow infiltration and gaseous exchanges, retain nutrients needed for plant and micro-organism growth, keep soil from compacting, and retain carbon from the atmosphere.

Organic matter is lost through deforestation, erosion and tillage. Organic matter can be increased by cover cropping and reduced tillage.

Food web

The food web in healthy soil contains more biodiversity than any other place on earth. Uncounted—and many yet unknown—micro-organisms in the soil breaking down complex materials or eating other organisms convert nutrients to forms that make them available to plants. Thus, the micro-organisms help plants get what they need, and plants help micro-organisms get what they need. This food web supplies the nutrition needs of all plant life on earth.

Appendix IV: *Soil*

Carbon

While this food exchange is happening, carbon is being sequestered in the soil. Some researchers believe that this holding of carbon in soil may, to a small degree, help offset the atmospheric carbons that contribute to global warming.[1] One farmer estimates that his farming practices have, in 26 years, "taken around 10 tons of carbon from the atmosphere and added it to the top seven inches of soil on every acre."[2]

In Napa County hillsides, where some people farm on only 6 inches of topsoil, carbon is being retained by permanent, no-till cover crops—which also help reduce erosion and improve filtration.

Adapted from USDA[3] and local soil data sources

NOTES

[1] Brian Lavendel, "Carbon in Our Soil: Building carbon in agricultural soils may offer a partial, short-term solution to increased carbon dioxide in the earth's atmosphere and global climate change." *Conservation Voices,* August-Sept. 2000, pp. 12–15.

[2] *Ibid.*

[3] Soil Quality Information series published by the National Soil Survey in cooperation with the Soil Quality Institute, NRCS, USDA and the National Soil Tilth Laboratory, Agricultural Research Service, USDA, Washington, D.C., 1995-1998.

Healthy or poor soil—you make the call

Napa County soils are considered rich and wonderful for winegrapes. The tremendous diversity of soils and microclimates contributes to the interesting and complex flavors expressed in the fruit and in the wines of Napa Valley. But each soil type in the county, from its slope and depth to its individual chemistry, needs to be viewed individually. Sometimes rich wonderful soil is prone to flooding or has drainage problems. Frequently a very thin soil on a steep slope is highly vulnerable to erosion. Soil quality can be assessed through observation and testing. Soil experts have focused on more than a dozen concerns that can be addressed by soil quality:

Soil quality issues

Loss of soil material by erosion
Deposition of sediment by wind or floodwaters
Compaction of layers near the surface
Soil aggregation at the surface
Infiltration reduction
Crusting of the soil surface
Nutrient loss or imbalance
Pesticide carryover
Buildup of salts
Change in pH to an unfavorable range
Loss of organic matter
Reduced biological activity and poor residue breakdown
Infestation by weeds or pathogens
Excessive wetness

Monitoring soil health—managers, choose your indicators

To take stock of soil quality and monitor changes, pick measurable properties that best reflect the soil function priority—at some sites the most important soil function may be vine health, and at another site the major soil function might be erosion control.

Your indicators will fall into any of four categories: visual, physical, chemical and biological (see chart). No single indicator can determine soil quality, and one measurement will not do the trick; indicators should be observed between one year and 10 years so that changes and trends can be observed and evaluated.

Indicators:

Visual

Measured by: eye or by camera

Examples: subsoil exposure, soil color changes, ephemeral gullies and blowing soil

Can indicate: changes or threats to soil quality

Physical

Measured by: visual and lab analysis of topsoil

Examples: topsoil depth, bulk density, porosity, aggregate stability, texture, crusting and compaction

Can indicate: limitations to root growth, seedling emergence or infiltration of water within soil profile

Chemical

Measured by: chemical analysis

Examples: analyses of pH, salinity, organic matter, phosphorous concentrations, cation-exchange capacity, nutrient cycling and contaminants (such as heavy metals, radioactive compounds and others)

Can indicate: soil-plant relations, water quality, buffering capacities, availability of nutrients, mobility of contaminants and crust formation

Biological

Measured by: respiration, microscopic (and other) analysis

Examples: presence, activity and byproducts of micro- and macro-organisms, organism populations, decomposition rates and pathogen populations

Can indicate: biological activity and soil aggregate stability

Appendix IV: *Soil*

Napa Valley Soils

Parent material is the term used to describe the mineral origins of a soil. This list shows the parent material of Napa County soils occurring on the hillsides and on the Valley floor.

Hillside soil parent material		Valley soil parent material
Oceanic Sediment & Rocks	**Sonoma Volcanics**	**Recent Alluvium**
OLD →		**YOUNG**
Bressa-Dribble complex	Aiken loam	Bale
Contra Costa	Boomer loam	Clear Lake
Diablo Clay	Boomer gravelly loam	Cole
Fagan clay loam	Forward gravelly loam	Coombs
Felton gravelly loam	Kidd loam	Cortina
Henneche gravelly loam	Guenoc	Egbert
Lodo	Hambright	Haire
Maymen		Maxwell
Los Gatos		Perkins
Millsholm		Pleasanton
Montara		Reyes
Sobrante		Riverwash
		Tehama
		Yolo

◄ Grower Al Buckland beside a Carneros vineyard soil pit. These pits allow vineyard managers to assess soil physical characteristics and other factors affecting vine growth and land management.

Soil-Vegetation Associations

Typically, certain soils are associated with vegetation that grows in them. This list shows common soil-vegetation associations found on Napa County lands.

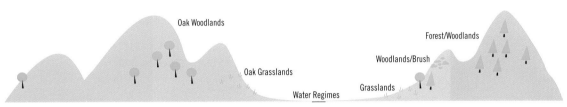

▲ This stylized illustration suggests a cross section of the Napa Valley and the general soil groupings as they commonly appear in different sites.

Forest/Woodlands

Aiken
Boomer
Felta
Felton
Forward

Brush

Kidd
Los Gatos
Maymen

Grasslands

Fagan
Diablo
Haire
Maxwell
Millsholm
Pleasanton

Oak-Grasslands*

Bale
Bressa
Clear Lake
Cole
Contra Costa
Coombs
Dibble
Egbert
Guenoc
Perkins
Sobrante
Tehama
Yolo

*the formal term is oak savannah

Water Regimes

Cortina
Reyes
Riverwash

Woodlands/Brush

Henneke
Lodo

Appendix Four: Soil

Appendix IV: *Soil*

General Soil Map, Napa County, California
U.S. Department of Agriculture Soil Conservation Service
University of California Agricultural Experiment Station

Soil Legend*

1–4: Well drained to poorly drained, nearly level to moderately steep soils on alluvial fans, on flood plains, in basins, on tidal flats, and on terraces.

5–11: Excessively drained to well drained, gently sloping to very steep soils on uplands

Terms for texture refer to the surface layer. (Compiled 1977)

1 **Bale-Cole-Yolo:** Nearly level to gently sloping, well drained and somewhat poorly drained loams, silt loams, and clay loams on flood plains, alluvial fans, and terraces

2 **Tehama:** Nearly level to gently sloping, well drained silt loams on flood plains and alluvial fans

3 **Reyes-Clear Lake:** Nearly level, poorly drained silty clay loams and clays on tidal flats, in basins, and on basin rims

4 **Haire-Coombs:** Nearly level to moderately steep, moderately well drained and well drained gravelly loams, loams, and clay loams on terraces

5 **Bressa-Dibble-Sobrante:** Moderately sloping to very steep, well drained loams, silt loams, and silty clay loams on uplands

6 **Henneke-Montara:** Moderately sloping to very steep, excessively drained and well drained gravelly loams and clay loams on uplands

7 **Maymen-Lodo-Felton:** Steep to very steep, somewhat excessively drained and well drained gravelly loams and loams on uplands

8 **Rock Outcrop-Kidd-Hambright:** Rock outcrop and gently sloping to very steep, well drained very stony loams and loams on uplands

9 **Forward-Boomer-Felta:** Gently sloping to very steep, well drained loams, gravelly loams, and very gravelly loams on uplands

10 **Forward-Aiken:** Gently sloping to steep, well drained gravelly loams and loams on uplands

11 **Fagan-Millsholm:** Moderately sloping to very steep, well drained loams and clay loams on uplands

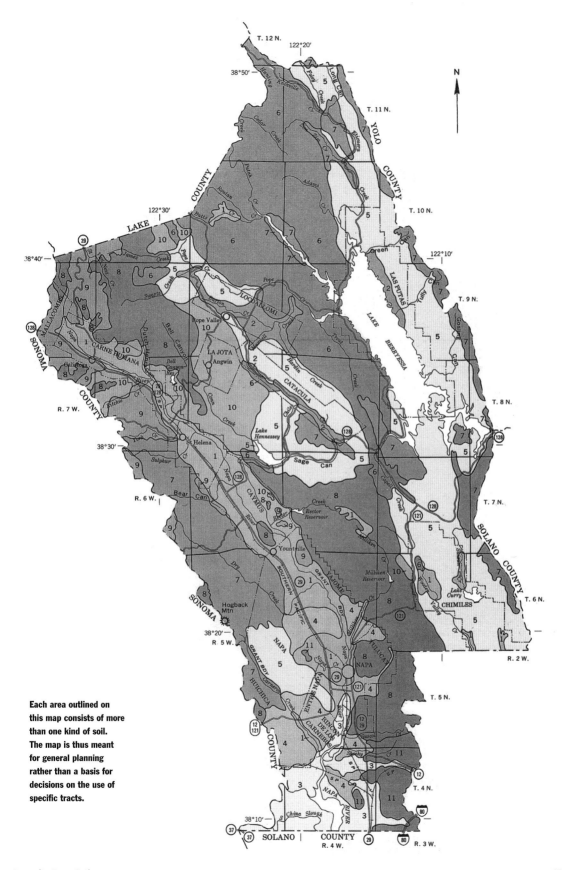

Each area outlined on this map consists of more than one kind of soil. The map is thus meant for general planning rather than a basis for decisions on the use of specific tracts.

Appendix Four: Soil

Appendix V: *Resources*

Napa Sustainable Winegrowing Group
Contact: Astrid Bock-Foster
(707) 252-4188
www.nswg.org

Napa County Resource Conservation District (RCD) and the Natural Resources Conservation Service (NRCS)
(707) 252-4188
www.naparcd.org

Napa County Agricultural Commissioner
(707) 253-4357

University of California Cooperative Extension
(707) 253-4221

California Department of Fish and Game
(707) 944-5500

Habitat for Hooters
Owl box sales benefit local non-profit groups.
Contact Janet Barth
(707) 224-3464

Napa River Watershed Owner's Manual: An Integrated Resource Management Plan
Technical and Educational Advisory Committees (1994). Napa County Resource Conservation District.

Integrated Pest Management: A Field Handbook for Napa County.
1st ed. The Napa Sustainable Winegrowing Group. 29 pp. (1997).

The North Coast Pierce's Disease Task Force Riparian Management Handbook is available at:
www.cnr.berkeley.edu/xylella/north/info.htm

Riparian Vegetation Management for Pierce's Disease in North Coast California Vineyards.
(Information Manual) Developed by the Pierce's Disease/Riparian Habitat Work Group (2000)

Cover Cropping in Vineyards: A Grower's Handbook
Eds. Chuck A. Ingels, Robert L. Bugg, Glenn T. McGourty and L. Peter Christensen 162 pp. (1998) University of California Division of Agricultural Resources Publication 3338
(800) 994-8849

Selected Cover Crop Seed Mixes for Napa County Grape Growers
Phillip Blake. NRCS District Conservationist, 2 pp. USDA Natural Resources Conservation Service (707) 252-4188

Know Your Natives: A Pictorial Guide to California Native Grasses
Jeanette Wrysinski, Yolo County Resource Conservation District (2000): Woodland.
(530) 662-2037, ext. 3

Vineyards in an Oak Landscape
Adina M. Merenlender and Julia Crawford, UC Berkeley
15 pp. (1998)
University of California Division of Agricultural Resources Publication 21577
(800) 994-8849

Farming for Wildlife: Voluntary Practices for Attracting Wildlife to Your Farm
Jeanne Clark and Glenn Rollins, eds. State, federal and private agencies.

Conservation Buffers to Reduce Pesticide Losses, March 2000 21 pp. USDA NRCS Fort Worth, Texas.

Conservation Buffers Work… Economically and Environmentally
USDA Program Aid 1615 Revised Sept. 2000, 16 pp. (folded)

Intercropping Principles and Production Practices
ATTRA Agronomy Systems Guide 15 pp. (1998) Appropriate Technology Transfer for Rural Areas (800) 346-9140

Acorn to Oak: A Guide to Planting and Establishing Native Oaks.
Circuit Riders Productions, Inc., Windsor. (1989).

Soil information

The soil food web: *www.soilfoodweb.com*

Soil information: *www.liveinc.org*

Soil biology: *www.statlab.iastate.edu/survey/sqi/soilbiology.htm*

Napa soil profiles, from surveys completed in 1978, are featured on a web site that includes historic information on soils and land use, soil profiles, soil features and engineering classifications, soil use and management, climatic data and countywide maps: *www.ca.nrcs.usda/mlra.napass/napass.html*

Weather and water

For current weather forecasts from the National Weather Service: *http://weather.gov*.

National Weather Service page: *www.wrh.noaa.gov*.

Historical and current rain and stream flow data from USGS and Department of Water Resources gauges: *http://cdec.water.ca.gov*.

You will need to know (or be prepared to look up online) the station identifications: Napa River at Napa (USGS **stream flow**) NAP; Napa River near St Helena (USGS **stream flow**) STH; Angwin (Department of Water Resources **rain**) ANG; St Helena (Department of Water Resources **rain**) SH4; Atlas Peak (Department of Water Resources **rain**) ATL.

More for Napa County: ALERT system of rain and stream level maintained by Napa County and the City of Napa. For ALERT systems gauge addresses in the San Francisco Bay Area: *www.wrh.noaa.gov/afos/sfo/rr1/sforr1sfo*. (Please note that current data sources are for provisional data.)

Bibliography

"2001 Soil Planning Guide." Natural Resources Conservation Service and Soil Science Society of America, 2000.

Adam, Katherine. "Suppliers of Organic and/or Non-GE Seeds and Plants." Fayetteville, Ark.: Appropriate Technology Transfer for Rural Areas, 2001.

Ahrens, W.H., ed. "Herbicide Handbook." 7th edition. Lawrence, Kan.: Weed Science Society of America, 1994.

Blake, Phillip. "Erosion Control Design in Hillside Vineyards." Napa: Natural Resources Conservation Service, 2001.

Blake, Phillip. "Going Native with Vineyard Cover Crops." *Napa Sustainable Winegrowing Group Newsletter* vol. 1, no. 1 (1998): 3–5.

Blake, Phillip. "Selected Cover Crop Seed Mixes for Napa County Grape Growers." Napa: Natural Resources Conservation Service.

Broome, Janet C. "California Winegrape Pest Profile, Pesticide Use, and Research Needs Under 1996 Food Quality Protection Act." *Sustainable Agriculture* vol. 12, no. 1 (2000): 12–17.

Broome, Janet C., Lisa C. Scott and Bonnie Hoffman. *Grower's Guide to Environmental Regulations & Vineyard Development.* Davis: Sustainable Agriculture Research and Education Program, Division of Agriculture and Natural Resources, University of California and Sacramento: California Association of Winegrape Growers, 2000.

Bryant, Dan. "Cover Crop Tames Vine Vigor." *California-Arizona Farm Press,* May 9, 1981.

Bugg, Robert L. and Mark Van Horn. "Ecological Soil Management and Soil Fauna: Best Practices in California Vineyards." *ASVO Viticulture Seminar: Viticultural Best Practice.* Adelaide, South Australia: Australian Society of Viticulture and Oenology, 1998.

Bugg, Robert L., et al. "Comparison of 32 Cover Crops in an Organic Vineyard on the North Coast of California." *Biological Agriculture and Horticulture* vol. 13 (1996): 63–81.

Bugg, Robert L., Cynthia S. Brown, and John H. Anderson. "Restoring Native Grasses to Rural Roadsides in the Sacramento Valley of California: Establishment and Evaluation." *Restoration Ecology* vol. 5, no. 3 (September 1997): 214–228.

Carpenter, E. and S.W. Cosby. "Soil Survey of the Napa Area, California." Washington, D.C.: USDA Bureau of Chemistry and Soils, 1938.

Castelle, A.J, A.W. Johnson and C. Conolly. "Wetland and Stream Buffer Size Requirements—A Review." *Journal of Environmental Quality* vol. 23 (September–October 1994): 878–882.

Chadwick, Anne. "The Winegrape Guidebook for Establishing Good Neighbor and Community Relations." Sacramento: California Association of Winegrape Growers, 2001.

Christensen, Damaris. "The World of Wine: Can Chemical Analysis Confirm a Wine's Authenticity?" *Science News* vol. 157 (January 1, 2000): 12–13.

Clark, J., et al. *Farming for Wildlife: Voluntary Practices for Attracting Wildlife to Your Farm.* Sacramento: 1996.

Clary, Patty. "Time for a Change: Pesticides and Wine Grapes in Sonoma and Napa Counties, California." Arcata: Californians for Alternatives to Toxics, 1997.

Clausen, J.C. et al. "Water Quality Changes from Riparian Buffer Restoration in Connecticut." *Journal of Environmental Quality* vol. 29 (November–December 2000): 1751–1761.

Coalition for Urban/Rural Environmental Stewardship and California Sulfur Task Force. "Sulfur Best Application Practices: Managing Sulfur Applications Near Sensitive Areas." Sacramento: CURES, 2001.

Committee to Assess the Scientific Basis of the Total Maximum Daily Load Approach to Water Pollution Reduction. "Assessing the TMDL Approach to Water Quality Management." Washington, D.C.: National Academy Press, 2001.

Cox, Jeff. "Organic Winegrowing Goes Mainstream." *The Wine News,* August/September 2000.

Diver, Steve. "Alternative Soil Testing Laboratories." Fayetteville, Ark.: Appropriate Technology Transfer for Rural Areas, 1998.

Dagget, Dan. *Beyond the Rangeland Conflict: Toward a West That Works.* Flagstaff, Ariz: The Grand Canyon Trust and Layton, Utah: Gibbs Smith, 1995.

Department of Pesticide Regulation. "Pest Management Grants Program." Sacramento: California Environmental Protection Agency, Department of Pesticide Regulation.

Diane, Susan. "Sustainable Farming for Quality Grapes: Robert Sinskey Vineyards." *Practical Winery and Vineyard* vol. 22, no. 5 (2000): 44–53.

Dillaha, T.A., R.B. Renau and D. Lee Mostaghimi. "Vegetative Filter Strips for Agricultural Nonpoint Source Pollution Control." Transactions of the ASAE vol. 32, no. 2 (March–April 1989): 513–519.

Donaldson, Dean, et al. "Weed Control Influences Vineyard Minimum Temperatures." American Journal of Enology and Viticulture vol. 44, no. 4 (1993): 431–434.

Dyer, David A. "Soil Sequestration of Carbon and Biomass-to-Ethanol." Natural Resources Conservation Service, 2000: TN-Plant Materials-58.

Eisele, Volker. "Twenty-five Years of Farmland Protection in Napa County" in *California Farmland and Urban Pressures: Statewide Regional Pressures.* ed. Medvitz, Albert G., Alvin D. Sokolow and Cathy Lemp. Davis: Agricultural Issues Center, Division of Agriculture and Natural Resources, University of California, 1999: 103–124.

Flaherty, Donald L. et al., eds. *Grape Pest Management.* Oakland: Division of Agriculture and Natural Resources, University of California, 1992.

Flosi, Gary et al. "California Salmonid Stream Habitat Restoration Manual." 3rd edition. The Resources Agency, 1998.

Grace, Kirk. "Goals and Objectives in Establishing Cover Crops for Erosion Control." Speech delivered at Erosion Control Workshop and Field Day, Napa County, Aug. 10, 1999.

Hargrove, W.L., ed. *Cover Crops for Clean Water.* Ankeny, Iowa: Soil and Water Conservation Society, 1991.

Hilty, Jodi Ann. "Use of Riparian Corridors by Wildlife in the Oak Woodland Landscape." Berkeley: University of California, August 2001.

Holmes, Henry. *Roots of Change: Agriculture, Ecology and Health in California.* San Francisco: Funders Agriculture Working Group, 2001.

Howell, David and Jonathan R. Swinchatt. "A Discussion of Geology, Soils, Wines and History of the Napa Valley Region." California Geology, May/June 2000.

Ingels, C., et al. *Cover Cropping in Vineyards: A Grower's Handbook.* Oakland: Division of Agriculture and Natural Resources, University of California, 1998.

Johnston, W.E. and H.O. Carter. "Structural Adjustments, Resources, Global Economy to Challenge California Agriculture." California Agriculture, vol. 54, no. 4 (2000).

Kegley, Susan, Stephan Orme and Lars Neumesiter. *Hooked on Poison: Pesticide Use in California 1991–1998.* San Francisco: Pesticide Action Network, 2000.

Klonsky, Karen, et al. "Sample Costs to Produce Organic Wine Grapes in the North Coast with an Annually Sown Cover Crop." Davis: University of California Cooperative Extension, 1992.

Kodmur, Julie Ann. "Soils in the Napa Valley: Q & A with Phill Blake." Napa Valley Wine Library Report Winter 1997. St. Helena: Napa Valley Wine Library Association, 1997: 5.

Kuepper, George ed. "Organic Certification and the National Organic Program." Fayetteville, Ark.: Appropriate Technology Transfer for Rural Areas, 2001.

Larson, Gary. *There's a Hair in My Dirt!* New York: HarperCollins Publishers, 1998.

Lavendel, Brian. "Carbon in Our Soil." Conservation Voices, August/September 2000.

Lee, Kye-Han et al. "Multispecies Riparian Buffers Trap Sediment and Nutrients During Rainfall Simulations." Journal of Environmental Quality vol. 29 (July–August 2000): 1200–1205.

Mayer, Kenneth E. and William F. Laudenslayer, Jr. *A Guide to Wildlife Habitats of California.* Sacramento: California Department of Forestry and Fire Protection, 1988.

Merenlender, Adina M. "Mapping Vineyard Expansion Provides Information on Agriculture and the Environment." California Agriculture, May–June 2000.

Motto Kryla & Fisher. "Vineyard Economics: An MFK Research Report Including the Updated MFK Vineyard Cost Study." St. Helena: Motto Kryla & Fisher LLP, 2000.

Napa County Resource Conservation District. "Huichica Creek Watershed Natural Resource Protection and Enhancement Plan." Napa County Resource Conservation District, 1993.

Napa River Watershed Owner's Manual: An Integrated Resource Management Plan: Napa, Napa County Resource Conservation District.

Napa Sustainable Winegrowing Group. "Integrated Pest Management: Field Book for Napa County." Napa Sustainable Winegrowing Group, 1997.

Natural Resources Conservation Service. "Conservation Buffers to Reduce Pesticide Losses." Fort Worth: Natural Resources Conservation Service, 2000.

Natural Resources Conservation Service. "EQIP Final Ranking Report 1999: Napa River/Putah Creek Watershed: California Geographic Priority Area #17." Natural Resources Conservation Service, 1999.

Osborne, Lewis L. and David A. Kovacic. "Riparian Vegetated Buffer Strips in Water-Quality Restoration and Stream Management." Freshwater Biology vol. 29 (1993): 243–258.

Oster, Jim et al. "Using Napa Sanitation District Wastewater for Irrigation of Grapes in the Carneros Region." Napa: Napa Sanitation District and Carneros Recycled Water Task Force, 1993.

Sharpley, A.N., et al. "Agricultural Phosphorous and Eutrophication." U.S. Department of Agriculture Agricultural Research Service, 1999: ARS-149.

Soil Conservation Service River Basin Planning Staff. "Hillside Vineyards Unit Redwood Empire Chapter Target Area: Napa and Sonoma Counties." USDA SCS, 1985.

Spence, Brian C. et al. "An Ecosystem Approach to Salmonid Conservation." Management Technology, 1996 (TR 4501-96-6057)

State Water Resources Control Board. "1990 Water Quality Assessment." Sacramento: SWRCB, Division of Water Quality, 1990.

Steiner, Rudolf. "Spiritual Foundations for the Renewal of Agriculture." trans. Catherine E. Creeger and Malcolm Gardner; ed. Malcolm Gardner. Kimberton, Pa.: Bio-Dynamic Farming and Gardening Association, 1993.

Stockwin, Will. "No-Till Experiments Pay Off." Fruit Grower, May 1988.

Sullivan, Preston. "Intercropping Principles and Production Practices." Fayetteville, Ark.: Appropriate Technology Transfer for Rural Areas, 1998.

Trumbo, Joel, Rob Forster and Tedd Yargeau, "California Wildlife and Pesticides." California Department of Fish and Game, Pesticide Investigations Unit, 1997.

U.S. Department of Agriculture, "Conservation Buffers Work…Economically and Environmentally." Program Aid 1615. Washington, D.C.: U.S. Department of Agriculture, revised September 2000.

Varela, Lucia G., *Pierce's Disease in the North Coast.* (Ed. Alexander Purcell, UC Berkeley and Rhonda J. Smith, Viticulture Advisor, Sonoma County). University of California Cooperative Extension & Statewide IPM Project.

Whitmer, Dave. "Historical Evolution of Agriculture in the United States" presented at *"Reunion Technique,"* Domaine Chandon, June 17, 1998.

Winkler, A.J., J.A. Cook, et al. *General Viticulture.* Berkeley and Los Angeles: University of California Press, 1974

York, Alan. "The Nuts and Bolts of Biodynamics: Balancing Resources Efficiently & Creating a Self-Sustaining System." Acres USA, April 1997.

▲ Community water demands were fewer in 1916. Horses stop for a drink on Spring Mountain Road, en route to market in St. Helena. (Photo courtesy of Tom Wilson.)

Index

Agricultural Preserve 73, 81
American Canyon 54
Araujo, Bart 126, 127
Araujo, Daphne 126, 127
Araujo Estate Wines 126
Armillaria root rot 82
Bambi syndrome 77
Bazan, Mario 140
Bell Canyon Reservoir 36
Beringer Vineyards 109
Beringer Wine Estates 137
Berkowitz, Zach 94, 95, 100, 101, 102
bindweed 66, 67
biodiversity 23, 76, 77, 78, 80, 83, 94, 96, 106, 113, 119,
biodiversity linkages checklist 80
biodynamic 126, 127, 128, 129
bird boxes 112

Blake, Phill 34, 36, 39, 160
Bluebird Recovery Program 112
blue-green sharpshooter 108
Bon Terra Vineyards 105
Bonotto, Vince 149, 150, 151
Bothe State Park 112
botrytis bunch rot 101
Buckmann, Allan 79
Bugg, Bob 91
Calderon, Angel 141
CALFED 46
California Association of Winegrape Growers 135
California Audubon Society 112
California Certified Organic Farmers 123
California Department of Food and Agriculture 123
California Environmental Quality Act 39
California freshwater shrimp 76, 149, 150, 151

California Grapevine Nursery 137
Calistoga 44, 48, 53, 141
Camera, Richard 126
camps, labor or farm worker 136, 137, 138, 141
Cantisano, Amigo Bob 120
Cardenas, Ramon 141
Carneros 29, 44, 47, 64, 65, 71, 107, 149
Carneros Creek 45, 46, 48, 75, 76, 144
Chappellet 88
Chiles Creek 90, 92
Chiles Valley 90
Christian Brothers 88
Clean Water Act 52
compost 107, 120, 121
Conn Creek 24, 108
conservation easements 81
conservation regulations 31, 34, 36

cover crops 68, 71, 87, 88, 90, 92, 93, 94, 95, 120, 121, 149, 100, 101, 102, 106, 111

cuadrillas 141

de Leuze Family Vineyards 107

de Leuze, Norman 107, 110

deficit irrigation 49

Del Bondio, Jim 120

Department of Fish and Game 74, 79, 109

Department of Pesticide Regulations 135

Devaux 149

diuron 62, 63

docks 67

Domaine Chandon 67, 68, 100, 101

Dry Creek 144

dry farming 48, 53

economic threshold 111

Eisele, Liesel 91

Eisele, Volker 90, 91, 92

Eisele Vineyards (Araujo) 126

Environmental Protection Agency 145

eutypa dieback 104

fencing 79, 83

Food Quality Protection Act 62

Friends of the Napa River 45

Frog's Leap Winery and Vineyards 79, 130

Galleron Ranch 79

Gardner, Shari 23

Garlock, DeWitt 88, 93

Garnett Creek 144

Geitner, Rex 33, 34, 35

glassy-winged sharpshooter 108, 138

glyphosate 63, 65, 67

Grace, Kirk 95, 96, 110, 111

Graves, David 46, 47

Green, Ray 123

Grigsby, Eric 70

groundwater 23, 43, 46, 51, 54, 56, 57, 151

Habitat for Hooters 113

Healy, Tim 54

Hess Collection Winery, The 126

hillside ordinance 31, 32, 39

Hilty, Jodi 78

Hoffnagle, Jon 81

Hopper Creek 144

Hoxsey, Andy 120, 121, 137

Hudson Vineyards 75, 77

Hudson, Lee 75, 76, 77, 144

Huichica Creek 48, 75, 144, 149, 150, 151

indicator species 77

insectories 80

integrated pest management 103, 104, 110, 111

irrigation 49, 53, 54

Johnson, Drew 108, 109

Johnson-Williams, Julie 70

Joseph Phelps Vineyards 128

Kanagy, Jon 66

Klug, Mitchell 64, 665, 53

K-selection species 66

Land Trust of Napa County 81

Larkmead Vineyards 106

leafhoppers 101, 111, 113

Lodi Woodbridge Winegrape Commission 145

Maher, Kelly 67, 68

McElroy, Brian 126

McGourty, Glenn 111, 112

Melgoza, Sergio 70

Mendocino County UC Extension 111

Merenlender, Adina 82

methyl bromide 142

Mexico 128, 134, 137, 138

Milliken, Beth 146, 147

Mochizuki, Martin 49

Mondavi, Robert 137

monoculture 105

Montes, Bullmaro 128, 129

Moore, Dennis 88, 93

morning glory 67

Mt. Veeder 45, 74

Mumm Napa 149

Napa 54, 134

Napa County Flood Control District 108, 109

Napa County Resource Conservation District (*see* Resource Conservation District)

Napa River 23, 24, 26, 79, 144

Napa River watershed 23, 24, 26, 45, 48, 50, 51

Napa Sanitation District 54

Napa Sustainable Winegrowing Group 58, 143, 145

Napa Wine Company 120

Napolitano, Mike 50, 51, 52

Natural Resources Conservation Service 32, 36, 37, 39, 48, 91, 108, 151

neutron probe 49

Niebaum-Coppola Winery 137

nitrogen fertilizers 94, 121

Nord Coast Vineyard Service 66, 142

Not the Silver Bullet Committee 109

Oak Knoll 53

oak root fungus 82

Oakville 120, 136

organic 90, 105, 119, 120, 122, 123, 125, 126, 130, 131

organic certification 123, 130

Organic Materials Review Institute 122

Oropeza, Reuben 137

owl boxes 112, 113

Pessereau, Phillippe 128, 129

Pesticide Action Network North America 62

pesticides 53, 57, 58, 82, 84, 93, 99, 100, 103, 111, 112, 119, 122, 124, 135, 145

phylloxera 107, 126, 139

Pierce's disease 27, 28, 46, 64, 108, 109

powdery mildew 101, 104

pressure bomb 49

Princep® 62

rainfall 43, 44

Ramirez, Francisco 140, 141

recycled water 54

red-root pigweed 66

Redwood Creek 144

Regional Water Quality Control Board 45, 50

Resource Conservation District 37, 39, 46, 47, 48, 84, 143, 144, 151

riparian corridors 27, 75, 78, 83

Robert Mondavi Winery 93, 64

Robert Sinskey Vineyards 95, 110

Rossi, Louise 119, 122

RoundUp® 63, 65, 67, 70, 126

R-selection species 66

Rutherford 79, 130, 136

Saintsbury Winery 46

Salvador Channel 144

St. Helena 48, 135, 138, 146

Scaggs, Boz 125

Scow, Kate 93

Selby Creek 106

setbacks 27, 28, 75, 149

Sharp, Leigh 144

Sierra Club 39

simazine 62, 63

Soil Conservation Service 32, 88

soil compaction 53, 94, 96

soil biodiversity 93, 121

soil health 28, 93, 121

soil microbial communities 71, 93, 95

Solari family 137

Sonoma County Grapegrowers Association 145

sorrel 66

spider mites 104

spotted owl 74, 77

Spottswoode Vineyard and Winery 135, 146, 147

Spring Creek 147

Spring Mountain Vineyards 33, 35

Stag's Leap 64

steelhead 45, 51, 151

Steiner, Dave 34, 37, 38, 39

Steiner, Rudolph 128, 129

Sterling 88

stewardship groups 28, 46, 48, 52, 76, 144

stream flow 23, 28, 48, 50, 144

sulfur 115, 123

Sulphur Creek 48, 144, 147

Syncaris pacifica 76, 149, 150, 151

terroir 28

thistle 67, 126

Thrupp, Ann 145

TMDL 45, 50, 51, 52, 55

Tofanelli Vineyards 48, 124

Tofanelli, Pauline 124

Tofanelli, Vince 124

Toth, Sylvia 34

US Geological Survey 46

UC Berkeley 78, 109

UC Berkeley Cooperative Extension 82

UC Cooperative Extension 82, 109

UC Davis 49, 88, 91, 93

Veeder Summit Ranch 126

vernal pools 83, 84

Volker Eisele Family Estates 90

Walsh, Russ 36

Walsh Vineyard Management Services 49

water budget 44, 48, 55

wetlands 74, 83, 84

wildlife habitat 73, 74

Williams, John 79, 130, 131

Williams Ranch 70

Wood family 136

woodland habitat 23, 77, 82

Wooster, Ted 74

Yountville 53, 120, 138

York Creek 34, 35

York Creek Vineyards 137

Yount Mill Vineyards 120, 121, 137

Zaccone, Dana 54

ZD Wines 107

Et justificata est sapientia ab omnibus filiis suis.